Cosmic Origins

M. Mitchell Waldrop

Cosmic Origins

Science's Long Quest to Understand How Our Universe Began

 Springer

M. Mitchell Waldrop 🆔
Washington, DC, USA

ISBN 978-3-030-98216-4   ISBN 978-3-030-98214-0   (eBook)
https://doi.org/10.1007/978-3-030-98214-0

This Springer imprint is published by the registered company Springer Nature Switzerland AG
The registered company address is: Gewerbestrasse 11, 6330 Cham, Switzerland

To A.E.F.

Acknowledgements

The book you're now reading is a much-expanded version of a report originally commissioned by the John Templeton Foundation (JTF), and published online by the Foundational Questions Institute (FQXi). I highly encourage you to take a look at the original, which can be found both at https://tin yurl.com/8uent6u5, and on the JTF site: https://tinyurl.com/2p86wf9k. It was written as part of a larger JTF program to produce deep-dive reviews on four aspects of fundamental physics: cosmological origins (this one), time, emergence, and fine tuning. Each of these reviews was supposed to be as scientifically accurate as possible, complete with citations to the original research, yet so clear, engaging, and easy to read that even non-scientists could use it as a reference.

This was a tricky balance to pull off, to put it mildly. And to the extent that I succeeded, I had lots of help. Alexander Vilenkin and Andrei Linde took the time to help me understand some of the subtleties of cosmic inflation theory, and João Magueijo explained how certain "constants" of physics may actually have varied over time; please accept my deep appreciation. My thanks also go to Anthony Aguirre and David Sloan at FQXi, and to Thomas Burnett from the John Templeton Foundation, who not only conceived of this project and coordinated the JFT/FQXi reviews, but provided generous support while I was writing it.

Then there was my editor Zeeya Merali, whom I have known since I was *her* editor back when she was writing mind-blowing cosmology and physics features for *Nature*. In either role, she has been a joy to work with. And of course, it was Zeeya who forwarded the JFT/FQXi version to Angela Lahee, editor extraordinaire at Springer, and asked if they would be interested in expanding it into a full-fledged book. To my delight, they said yes—with the result that you now hold. So to Angela and her team, a heart-felt thank you.

Finally, none of this would have been possible at all without the love, devotion, support—and yes, patience—of my wife Amy. Thank you, thank you, and thank you.

January 2022

Contents

About the Author

M. Mitchell Waldrop is a freelance writer and editor. He earned a Ph.D. in elementary particle physics at the University of Wisconsin-Madison in 1975, and a Master's in journalism at Wisconsin in 1977. From 1977 to 1980 he was a writer and West Coast bureau chief for Chemical and Engineering News. From 1980 to 1991 he was a senior writer at Science magazine, where he covered physics, space, astronomy, computer science, artificial intelligence, molecular biology, psychology, and neuroscience. He was a freelance writer from 1991 to 2003 and from 2007 to 2008; in between he worked in media affairs for the National Science Foundation from 2003 to 2006. He was the editorial page editor at Nature magazine from 2008 to 2010, and a features editor at Nature until 2016. He is the author of Man-Made Minds (Walker, 1987), a book about artificial intelligence; Complexity (Simon & Schuster, 1992), a book about the Santa Fe Institute and the new sciences of complexity; and The Dream Machine (Viking, 2001), a book about the history of computing. He lives in Washington, D.C. with his wife, Amy E. Friedlander.

1

Introduction

They're some of the oldest questions that human beings have ever asked—renewed again and again by every child who looks up in wonder at the sun, the moon, the stars, and planets: What *are* they? Why do they move and change the way they do? Where do they all come from? And where do *we* come from?

These questions are so fundamental that every culture and every religion provides answers—often in the form of origin stories that illuminate equally fundamental questions about the group's identity, worldview, values, and purpose. Who are we? How should we live our lives? What is our role in the cosmos [1]?

Those origin stories are a fascinating study in themselves, ranging from the Acoma Indians' tale of humankind's birth from the womb of the Earth, to the Hebrews' story of God's creating the cosmos from nothing, to the Zulu tale of a hero who created mountains, cattle, people and everything else from the reeds [2]. But science doesn't try to answer the cultural questions these stories pose—or rather, it doesn't try to answer them directly. Instead, science focuses on the kind of factual questions raised in our first paragraph: mysteries that can be addressed by methods rooted in reason, experiment, and meticulous observation.

Yet, as this book will explore, that focus has guided scientists to a cosmic story that is far stranger than our ancestors could have imagined.

© The Author(s), under exclusive license to Springer Nature
Switzerland AG 2022
M. M. Waldrop, *Cosmic Origins*,
https://doi.org/10.1007/978-3-030-98214-0_1

1.1 Four Radical Shifts in Perspective

Our long journey toward this 21st-century story has spanned millennia. But more than that, it has required (at least) four radical shifts in perspective.

The Planets are Other Worlds, and the Stars are Other Suns

To readers who have grown up on science fiction tales depicting other worlds in other galaxies, this answer to the "What are they?" question might seem obvious. In fact, it required a massive upheaval in Western thought, which began in 1543 when the Polish astronomer Nicolaus Copernicus published a sun-centered model of the universe that he had been working on for the past three decades [3]. Prior to that, just about everyone had believed the evidence of their senses—that the Earth was solid and immovable—and assumed that the universe revolved around us.

Copernicus' primary motivation for challenging this assumption was mathematical beauty: He realized that the complex, looping movements of the celestial lights known as *planets* (from the Greek word for "wanderer") would make far more sense if they were actually just simple circular motions around the Sun—but circular motions as seen from a moving vantage point (the Earth) that was whirling around the Sun in its own orbit.

Yet this mathematical exercise was also a demotion of humankind's home from the cosmic center, a move that upended entrenched ideas about physics, morality, and even the God-man relationship—not to mention our own sense of self-importance [3]. This may have been why Copernicus published his theory only when he was near death, and only after much persuasion. It was definitely why the Catholic Church would ban Copernicus' book outright in 1616 [4].

Still, the evidence for this Copernican picture continued to accumulate, most famously when the Italian physicist Galileo Galilei built one of the first telescopes and pointed it at the heavens. He published his observations in 1610 as a short pamphlet whose Latin title is generally translated as *The Starry Messenger* [5]. What he saw in the sky—including mountains on our moon, and four previously unknown moons orbiting Jupiter—proved that these celestial points of light were far more like our world than ever imagined.

The Heavens and the Earth are One, and Operate According to Natural Law

Again, this seems axiomatic in the age of interplanetary spacecraft. Yet for most of human history it seemed just as obvious that the celestial realm is profoundly different from the base matter here on the ground, and that every event—storms, childbirth, the coming of spring, victory in battle, everything—depends on the whim of the gods. It wasn't until the time of Aristotle that the ancient Greek philosophers began to think in terms of natural law: fixed rules that apply everywhere, to everything, at all times. And it wasn't until the 1600s that this notion was made mathematically rigorous by the English polymath Isaac Newton; his laws of motion and gravity gave his fellow natural philosophers a set of equations that governed both the orbit of the moon and the fall of an apple, and that could predict the motions of planets, satellites, comets, projectiles, and anything else in the cosmos.

The Universe is Very Large and Very Old

On an everyday human scale, a thousand kilometers is a long, long way and a thousand years is forever. But from the cosmic perspective, both are miniscule. The immense size of the universe was already implied by Copernicus' theory in the 1500s. If you were willing to believe that a tiny, reddish dot of light like Mars was in fact a world like our own, then you also had to believe that it was ridiculously far away. (The modern figure is 55 million to 400 million kilometers, depending on where Earth and Mars are in their orbits.) The fixed stars had to be much further still.

Even then, however, most natural philosophers continued to accept that the universe was a few thousand years old (roughly 6000, according to *Genesis*). Their thinking began to change only in the late 1700s, when pioneering geologists began to realize that the ancient rocks they saw in cliffs, quarries, and road cuts had been formed by erosion, sedimentation, volcanic activity, and the like—the same processes that are slowly and steadily reshaping the landscape today. But the critical word was "slowly": When those early scientists calculated how long it would take geologic processes operating at today's rates to create the existing landforms, they came up with estimates in the millions of years. So the Earth—and by extension, the heavens—must be at least that old. (The modern figure is 4.6 billion years for the Earth and everything else in the solar system, and 13.8 billion years for the entire universe.) Scientists henceforth had to deal with the dizzying reality of what one 19th-century natural philosopher called "the abyss of time" [6].

The Universe Started Small and Grew

According to the origin stories found in *Genesis* and many other traditional accounts, the world around us was formed pretty much as we see it now—plants, animals, mountains, oceans, all brought forth in a single act of

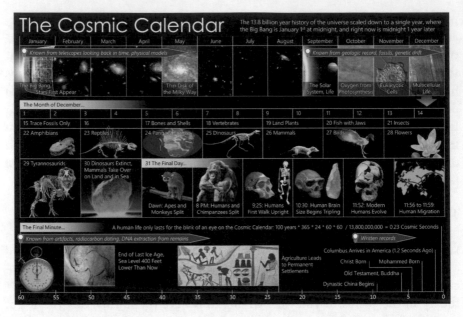

Fig. 1.1 The 13.8-billion-year lifetime of the universe mapped onto a single year. This one-year scale was popularized by the late astronomer Carl Sagan. (Credit: Efbrazil/Wikimedia Commons, CC BY-SA 3.0)

creation [7]. But according to the story that's been uncovered by science over the past two centuries or so, cosmic history is a long process of *becoming*: everything that we see today took shape according to natural law from much simpler beginnings.

The quest to understand how that happened—and what those cosmic beginnings might be—has defined much of 20th- and 21st-century astronomy and physics, and is the subject of this book (Fig. 1.1).

1.2   The Modern Shift

The cosmic story we've arrived at today is familiar enough. Our universe began with the "Big Bang," an event some 13.8 billion years ago in which space, time, matter, energy, light, and everything else came into being as an infinitesimal point of infinite temperature and density. And the universe has been expanding ever since, allowing the superhot energy of that initial point to cool and condense into electrons, protons, atoms, galaxies, stars, planets and eventually, us.

As we'll see, however, getting to this answer has required that scientists shift their perspective in ways that were at least as profound as any that came before—and just as difficult. Again and again, what are now considered to be foundational discoveries were met with indifference, incomprehension, or even hostility—and achieved widespread acceptance only after accumulating evidence made the new ideas impossible to ignore.

Chapter 2: The Expanding Universe

In Chap. 2 we review how this dynamic played out in the discovery of the first key piece of evidence for the Big Bang: the realization that just about all the galaxies in the universe are flying apart from one another, like sparks from some titanic explosion. This was a story that unfolded along two parallel tracks, as theorists and observers found themselves coming to the same conclusion only after working in near-total ignorance of one another.

The theoretical track began with Albert Einstein and his two theories of relativity. The 1905 version, now known as the *special* theory of relativity, showed that space and time are different aspects of a single, underlying unity: space–time. The 1915 version, known as the *general* theory of relativity, showed that space–time can bend, ripple, and curve—and that its curvature is the origin of the force we call gravity. In 1917, Einstein also laid the foundations for modern cosmology by pointing out that the equations of general relativity determine the shape and dynamics of the universe as a whole.

On the observational track, meanwhile, astronomers' understanding of the universe's true scale was undergoing a dramatic expansion of its own. They already knew that the visible stars are an astonishing distance away—so far that their light takes years to reach us. But in the 1910s, observers training ever more powerful telescopes on the sky discovered that these nearest-neighbor stars comprised only a tiny fraction of our Milky Way galaxy, which in turn proved to be an immense flattened disk many tens of thousands of light years across. Then in the 1920s, astronomers found that even this huge structure is just a dust mote on the cosmic scale. When they looked more closely at the mysterious pinwheels known as "spiral nebulae", the structures turned out to be star-filled galaxies just like ours, but located at distances measured in millions of light-years. And finally, at the dawn of the 1930s, astronomers realized that this vast cosmos isn't just big, but getting bigger. The universe—in keeping with Einstein's equations—is expanding quite literally.

Chapter 3: The Discovery of the Big Bang

In Chap. 3 we trace how astronomers and physicists confirmed that the universe did indeed begin in a cosmic fireball. Again, this conclusion did not come quickly or easily. The implications of cosmic expansion may seem obvious in retrospect. If the galaxies are flying apart now, after all, then they surely must have exploded outward from some much denser initial state somewhere in the distant past, billions of years ago. But few astronomers in the 1930s were comfortable with the idea of a cosmic beginning. And even fewer had time for speculations about an event that (they thought) could never be observed.

That attitude began to change only in the 1940s, when a handful of scientists realized that the new field of nuclear physics allowed them to observe the fireball indirectly, by calculating how thermonuclear reactions would have unfolded during the first few minutes of the universe. They found that the suite of chemical elements produced in those reactions would form a kind of fossil record of the event. And indeed, the abundances they calculated for the various isotopes of hydrogen and helium—the lightest, and by far the most abundant elements in the cosmos—turned out to match the observed values very closely.

This account of the observed abundances is now considered a second key piece of evidence for the Big Bang, after cosmic expansion. Yet even then, many scientists continued to be reluctant to consider origins–an attitude that was popularized for a while as the "steady-state" model, in which the universe had no beginning or end. But then in 1964, radio astronomers uncovered a third key piece of evidence: a faint whisper of low-energy radio waves now known as the Cosmic Microwave Background (CMB). These waves are essentially the Big Bang's afterglow, made up of photons that were emitted some 380,000 years after that initial cataclysm, when the superhot plasma it produced had cooled the point where electrons and protons could condense to form neutral hydrogen atoms and the universe became transparent.

The CMB not only made the Big Bang idea[1] almost inescapable, but it has proved to be our richest source of information about the very early universe— a subject we will return to again and again.

Chapter 4: Behind the Veil

This chapter recounts how our understanding of the Big Bang was enriched in the 1960s and 1970s by one of the most spectacularly successful achievements in modern physics: The development of a *standard model* that provided a

---

[1] It's worth noting that many scientists use the phrase "Big Bang" in different ways. Some apply it only to the initial singularity that's conjectured to occur at the very first instant of the universe, when the cosmos is compacted to a point of infinite temperature and density. But many others use the phrase as a shorthand for some or all of the expansion and condensation processes that led up to the formation of the CMB. In this book we will generally follow the second convention, but will always try to make the meaning clear in context.

comprehensive and concise theory of all the known elementary particles, as well as all the non-gravitational forces that act between them.

We'll also see how this work on the fundamental nature of matter ended up convincing most cosmologists that the Big Bang itself emerged from a prior event: an incomprehensibly brief interval of incomprehensibly rapid expansion that stretched cosmic space–time as taut as a hyper-inflated balloon. Only after that period of "inflation" would a multi-billion-light-year patch of space–time slow down and begin the comparatively tame expansion we see today. Not every cosmologist accepts this inflation idea. But the majority feel they have no choice. Without invoking some kind of ultrarapid cosmic expansion, they have a very hard time explaining certain key features of our present-day universe, such as the fact that it's big, old, and looks pretty much the same in every direction. An inflation-less origin would have been much more likely to produce a distorted, convoluted universe that collapsed back on itself very quickly.

Chapter 4 will conclude with a look at what happens when this inflation idea is followed to its logical, if exceeding weird, conclusion: If an inflating cosmos can produce one bubble of normal, non-inflating space–time—our universe—then it's hard to see what would stop it from producing a multitude of others. These bubbles of normal space–time would rarely, if ever, collide with one another. But each would be a universe in its own right—perhaps even with its own laws of physics—and they would collectively comprise a kind of cosmic foam known as the "multiverse."

Despite the inflation model's popularity, though, inflation's proponents have yet to explain when and how inflation itself got started. This has led various critics to propose alternative frameworks. We will look at several. But it may be that the question of inflation's beginning doesn't need an answer. As advocates often point out, the background inflation process may well be eternal. From that very grand perspective, the larger multiverse needn't have a beginning at all.

Chapter 5: The Dark Universe

Next, we look at how cosmologists are testing theories like inflation with ever more accurate surveys of the CMB and galactic distances; with the detection of space–time ripples known as gravitational waves; and with ever-larger computer simulations of complex processes such as galaxy formation. These high-precision observations have already revealed one astonishing fact, which is that most of our universe is utterly invisible.

The first component of this invisible sector is a haze of "dark matter" that is about five times as massive as all the visible stars and galaxies put together, and that doesn't interact with ordinary matter at all. No one knows for sure what dark matter is—a swarm of previously unknown elementary particles created in the Big Bang, maybe?—but it has an overwhelming gravitational influence on the visible stars, which was how it was discovered in the 1970s.

The other component of the invisible sector is even more mysterious. Known as "dark energy," it was discovered in the 1990s through high-precision studies of the cosmic expansion rate, and seems to be some kind of universal cosmic repulsion that is causing the expansion of our universe to slowly speed up.

Chapter 6: The Age of Precision Cosmology

The discovery of the dark universe has helped scientists to home in on a *concordance* model of cosmology—the combination of Big Bang, inflation, dark matter, and dark energy that together seem to provide the best available description for our cosmic origins.

Or maybe not. In this final chapter, we look at three nagging loose ends in the concordance model that might just point the way to what lies beyond. One that has increasingly made itself felt over the past decade is the so-called Hubble tension: a small, but worrisome, discrepancy between the cosmic expansion rate obtained from examining the cosmic microwave background radiation, and the rate obtained from measurements "nearby" galaxies only a few billion light years away. These two approaches have led to a billion-year discrepancy in our estimates of the time since the Big Bang. The question is whether this is the result of some calibration error, or is revealing something new and profound about our understanding of the universe.

Another loose end might be called the case of the missing WIMPs. If dark matter is indeed a swarm of weakly interacting, massive particles left over from the Big Bang, as most cosmologists expect, then…where are they? Why has every attempt at direct detection fallen short? Could the WIMP picture be utterly off base?

And finally there is perhaps the ultimate loose end: What banged? Or more precisely, how did the Big Bang, inflation, dark matter, dark energy and all the rest arise from natural law? The answers to this question will almost certainly have to wait for a fully satisfactory theory of quantum gravity. And that, in turn, will probably require some profound new ideas about the nature of space, time, and quantum reality itself.

But the good news is that the coming decade should bring a flood of fresh data from a new generation of ultra-high-precision space- and ground-based telescopes. Our understanding of cosmic origins is poised to become much richer, for sure—and perhaps be utterly transformed.

References

1. Leeming DA, Leeming MA (1996) A dictionary of creation myths (Oxford Reference S), 1st edn. Oxford University Press
2. Leeming DA (2010) Creation myths of the world: an Encyclopedia. ABC-CLIO
3. Kuhn TS (1957) The copernican revolution: planetary astronomy in the development of western thought. Harvard University Press
4. Heilbron JL (2010) Galileo. Oxford University Press, Oxford
5. Galilei G (1957) Discoveries and opinions of Galileo: including the Starry Messenger (1610), letter to the Grand Duchess Christina (1615), and excerpts from letters on Sunspots (1613), The Assayer (1623). Doubleday
6. Playfair J (1805) Hutton's Unconformity. Trans R Soc Edinb V (III)
7. Leonard SA, McClure M (2004) Myth and knowing: an introduction to world mythology. McGraw-Hill

2

The Expanding Universe

Compared to the ceaseless ebb and flow of events on Earth, the heavens have always seemed like the essence of permanence and stability. So when physicists and astronomers discovered during the early decades of the twentieth century that our universe is actually expanding, the revelation came as both an utter surprise and a profound shock. Even some of the greatest of them stoutly resisted the idea—until the evidence became overwhelming.

The story of how this happened unfolded along two parallel tracks, as theorists and observers working in near-total ignorance of one another found themselves arriving at exactly the same conclusion.

2.1 The Fabric of Space and Time

2.1.1 Maxwell's Conundrum and Einstein's Resolution

The theoretical track began in 1905, when a young German physicist named Albert Einstein was working as an examiner at the Swiss Patent Office in Bern, and found himself puzzling over the behavior of light.

Like many in his generation, the 25-year-old Einstein had been deeply influenced by the work of James Clerk Maxwell [1], a Scottish physicist who had mathematically demonstrated in the 1860s that electricity, magnetism, and light are not three different things. They are three different aspects of the *same* thing: a unity now called *electromagnetism*.

© The Author(s), under exclusive license to Springer Nature
Switzerland AG 2022
M. M. Waldrop, *Cosmic Origins*,
https://doi.org/10.1007/978-3-030-98214-0_2

Maxwell had laid out his theory, arguably the greatest achievement of nineteenth century physics, in two parts. In the first, published in 1861, he showed that everything experimenters had discovered about electricity and magnetism could be summarized in just a handful of equations [2].[1] Then in the second part, published four years later, Maxwell used those equations to show that oscillating electric and magnetic fields could reinforce each other and go rippling across the universe in the form of a wave. Maxwell's calculations also showed that the velocity of that wave would be equal to a certain combination of numerical constants, all of which had long since been determined by from experiments with electricity and magnetism in the laboratory. And when Maxwell plugged in the measured values of those constants, he found that his hypothetical electromagnetic waves would move at roughly 300,000 km per second—which just happened to be the empirically measured speed of a light beam.

Now, Maxwell and his contemporaries already knew that light was a wave of some kind; the British physician Thomas Young had demonstrated that fact decades earlier, in 1803—although no one had a clue what this wave consisted of. But Maxwell's calculations suggested an obvious conclusion: this electromagnetic ripple *was* light. And conversely, light was an electromagnetic ripple.

This conclusion was proved correct more than two decades later, when the German physicist Heinrich Hertz used an early version of what would now be called a radio transmitter to generate electromagnetic waves in his laboratory [3–5].

Among many other things, Maxwell's discovery hugely expanded our concept of what "light" could be (Fig. 2.1). In terms of wavelength, defined as the distance between one crest of a wave to the next, Young's experiments had already shown that visible light occupies a narrow band between 380 nm at the violet end of the rainbow and 740 nm at the red end. (In modern terms, that's the size of a large-ish virus.) Young and others had also shown that there exists an invisible form of *ultraviolet* radiation that has wavelengths shorter than violet, and an equally invisible form of *infrared* heat radiation lying at wavelengths longer than red [6]. But now, post-Maxwell, scientists could begin to think about and work with electromagnetic waves lying far beyond the visible spectrum. Going from the longer to shorter wavelengths and using the modern names, these waves include radio (wavelengths measured in meters or even kilometers), microwaves (centimeters), infrared, visible, ultraviolet, X-rays, and gamma rays.

---

[1] Maxwell's original papers had 20 equations. But in the more compact notation used in modern textbooks, the same physical principles can be expressed in just four equations, or even two.

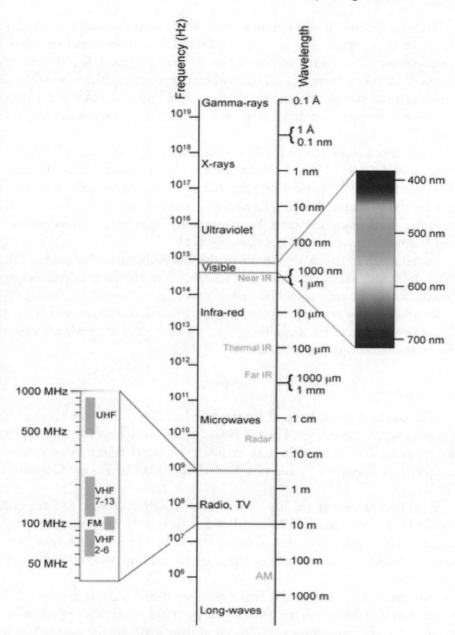

Fig. 2.1 Visible light turns out to occupy only a very small range of wavelengths. The full span includes everything from gamma radiation to the very longest radio waves. (Image credit: Penubag/Wikimedia Commons, CC BY-SA 2.5)

But what Einstein found puzzling about all this was Maxwell's prediction about the wave's speed. The velocity of light didn't seem to work like speeds do in everyday life. If you run after a bus, for example, the bus will seem to be moving slower relative to you, simply because you are catching up with it. And if you run alongside at exactly the same speed that the bus is moving, it will seem to be stationary from your perspective, and you can easily hop on board.

But even as a schoolboy in the 1890s, Einstein later wrote, he realized that things would get weird if he tried that same trick with a light beam. If he ran alongside at exactly the speed of light, then presumably he could look over and see the beam just hanging in mid-air—a set of oscillating electric and magnetic fields going nowhere. Yet a stationary light wave was something that Maxwell's equations did not allow for at all.

Einstein published his resolution to this conundrum on September 26, 1905. His paper's title, which in English reads "On the Electrodynamics of Moving Bodies," clearly shows his concern for Maxwell's work [7, English: 8

Relativity theory is renowned for Einstein's proof that mass, m, and energy, E, are two aspects of the same thing—a relation expressed in what is easily the most famous equation ever written:

$$E = mc^2 \tag{2.1}$$

(The constant c is shorthand for the speed of light.) But that idea didn't actually appear in the original paper. Einstein published it in November 1905 almost as an afterthought, in a short follow-on paper whose German title translates as, "Does the Inertia of a Body Depend on its Energy Content?" [8, 9].

What that first paper *did* have was an assertion that is simple to state, but radical in its implications: The speed of a light beam is the same for every observer. Never mind how that observer is moving—or for that matter, how the light-beam's source is moving—the beam will still zip past at 300,000 km per second.

This assumption solved Einstein's schoolboy conundrum at a stroke. No matter how fast he ran after a light beam, he could never catch up. And he could forget about running at the speed of light itself: By the same analysis that leads to $E = mc^2$, accelerating an object to light speed would require an infinite amount of energy.[2]

What Einstein *would* notice as he chased the light beam was a change in its wavelength: if he were running away from the source—or equivalently, if the source were moving away from him—the wave would appear to be stretched out and shifted toward the red end of the spectrum. Such a source is said to be *redshifted*, meaning that the light will look redder and redder as the source moves away faster and faster. Conversely, if Einstein were running toward the source, the waves would appear to compress and become "blue-shifted" toward the shorter-wavelength end of the spectrum. But either way, he would measure the actual speed of the wave as that same 300,000 km per second.[3]

The trick was to make this constant speed-of-light assumption work mathematically. Einstein devoted much of his paper to this—in the process, deriving all the famously weird consequences of relativity. If I see your rocket ship moving past me at some velocity, for example, then I will see you, your ship, and everything in it appear to contract in the direction of motion. (You, of course, would not notice any such contraction in yourself—instead, you would see me, my ship, and everything in it contract by the same amount.) Likewise, we would each see every clock in the other's ship slowing down, including our respective heartbeats and brain rhythms: time on the other ship would literally appear to be flowing at a different rate. And weirder still, if you perceived two events as happening simultaneously, I might very well see them as happening at different times—and vice versa.

As strange as these effects are, however, they all boil down to one deceptively simple fact: space and time are not separate things. They are two aspects of an underlying unity—aspects that are perceived differently by observers that are moving relative to one another. Thus the name, "theory of relativity."

It was a unification at least as profound as the one that Maxwell had achieved. And within a few years of Einstein's 1905 paper, physicists had begun referring to this underlying unity as *space–time*.

---

[2] The one exception is if an object has zero mass to start with—the best-known example being the *photon*, the particle of light that Einstein predicted in another 1905 paper. Such a massless particle has finite energy, but cannot move slower than the speed of light.

[3] As discussed later in this chapter, wavelength shift would prove pivotal in establishing that our universe is expanding.

2.1.2   Space, Time, and Gravity

Einstein took that unification to heart and made it the centerpiece of his long quest to find a more general version of relativity theory. He had felt from the beginning that his 1905 version, now called *special* relativity, was incomplete. It dealt with observers who were moving in a straight line at a constant speed, but said nothing about observers in, say, a rocket ship that was accelerating. It dealt with Maxwell's electromagnetic forces, but said nothing about the force of gravity.

After a decade of false starts and partial successes, Einstein finally got what he wanted—a conceptually elegant account of space, time, and gravity that dealt with all observers, no matter how they were moving. He presented his completed general theory of relativity at a conference talk on November 25, 1915 [10, English: 11], and then again in a detailed paper published the following year [12, English: 10

A crucial element in this success was Einstein's realization that space–time isn't something that's just *there*—a passive, rigid framework that exists only as a kind of stage for matter to do its thing. Space–time is dynamic. It can curve and ripple. It can expand and contract. It can even guide how particles move. In fact, said Einstein, that's what gravity *is*—not a force as we usually understand it, but a warping of space–time that's produced by a star or a planet's very presence, and that tries to make every other object move in that body's direction. The standard analogy is that the sun bends spacetime like a bowling bowl resting on a rubber sheet, while the planets that orbit the sun are just following the contours of the warped sheet (Fig. 2.2).

Testing this theory was tricky, though. For masses that are small on some cosmic scale, and for objects moving much, much slower than the speed of light—a description that fits pretty much everything in the solar system, including the sun and even the fastest comets and asteroids—the warping of space–time is infinitesimal, and the motions predicted by general relativity are virtually identical to those that Isaac Newton would have predicted in the 1600s. That's why we never see any kind of space–time distortion in our daily lives.

Still, Einstein came up with three observations in which the tiny effects of general relativity might be detectable. One, an infinitesimal slowdown in the vibration of atoms located deep in a gravitational field, was too subtle for instruments of the day; this prediction wouldn't be confirmed until the 1950s. But another effect—a tiny, but steady shift in the orbit of the fastest-moving planet, Mercury—had been known (and defied explanation) since the nineteenth century; general relativity fit the anomalous data almost exactly.

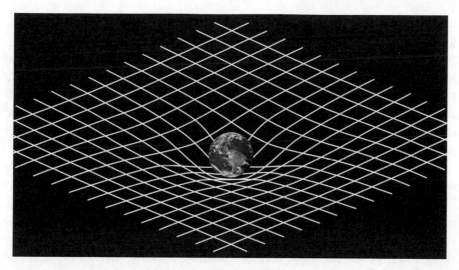

Fig. 2.2 Massive objects like planets curve spacetime, as shown in this artist's illustration of Einstein's conception. (Image credit: NASA's Imagine the Universe, Public Domain)

And Einstein's third prediction, a slight deflection of starlight passing close to the sun, was confirmed in spectacular fashion by British astronomers observing the total solar eclipse of May 29, 1919 [13, 14].

2.1.3 The Foundations of Cosmology

News of the eclipse discovery was front page news around the world in 1919, and made Einstein an overnight celebrity—a status that he alternately found amusing, baffling, and irritating. He avoided the distractions of fame as best he could, and instead tried to focus on understanding the implications of his theory. This work included a 1917 paper that arguably laid the foundations for modern cosmology, that would have immense implications for our understanding of space–time's origins—and that made Einstein himself quite uncomfortable [15].

In many ways this paper was the logical next step: having shown that the equations of general relativity described the motion of isolated stars and planets, Einstein found it natural to ask what they could tell us about the universe as a whole. He began by applying the equations to what seemed at the time like a common-sense approximation of the universe: an immense sphere filled with a uniform distribution of stars. This sphere would be so vast that its curvature would be invisible on the scale of the solar system, in the same way that the surface of the spherical Earth looks flat when we're standing on the ground. But on a scale of billions of light years, the curvature would be enough to make the universe spherical and finite.

Next, Einstein assumed that the universe as a whole was stable and unchanging. Again, this seemed like a pretty safe assumption in 1917: How else could the cosmos have lasted for millions or billions of years, as it obviously had? So Einstein searched for a solution to his equations that would confirm this picture, meaning that the solution would predict a cosmic sphere with a radius that stayed fixed, and would never vary with time.

But that's where things got disturbing. When Einstein did the math, the equations of general relativity kept insisting that the universe was anything but stable. No matter how his cosmic sphere started out—expanding, stationary, contracting—it would always end up collapsing to a point. The mutual attraction of all those stars would eventually win out and pull everything inwards, for the same reason that a fly ball will eventually return to earth: It can't just hang in mid-air forever with nothing to hold it up.

To Einstein, this result was unacceptable. The universe *had* to be stable. He felt this so strongly, in fact, that he ended up modifying general relativity itself to make it true, by adding a tiny *cosmological constant* to the equations. Like the overall cosmic curvature, he reasoned, this constant would be too small to detect on the scale of the solar system. But over cosmological distances, it would produce a tiny, but steady repulsion—just enough to counteract the mutual gravitation of the stars and prevent a cosmic collapse.

Adding this cosmological constant worked mathematically, and gave Einstein the cosmic stability that he felt was essential. But even so, it always stuck him as an ugly, ad-hoc way of solving the problem. He never liked it. Despite this, he continued to believe in the constant for nearly a decade and a half—even in the face of alternative theoretical proposals that suggested his picture was wrong [16].

In the early 1920s, for example, the Russian mathematician Alexander Friedmann generalized Einstein's cosmic model, and showed that a spherical universe was just one possibility among many others allowed by general relativity—with or without a cosmological constant [17, 18, English: 19]. Friedmann's solutions encompassed both expanding and contracting universes, and allowed for curvatures that were positive (Einstein's sphere), zero (infinite flat space), or negative (an infinite space shaped something like a saddle.) Today, those solutions hold a place of honor in every cosmology textbook. But at the time, Einstein just assumed that Friedmann had made a mathematical error. Even after he was convinced that the Russian's equations were correct, he dismissed the non-static results as "unphysical."

Fig. 2.3 The spiral nebula in the constellation of Andromeda is undeniably beautiful. But what is it? A nearby gas cloud? A newborn star? Another "island universe" like our own Milky Way? What? (Image Credit: David Dayag, CC BY-SA 4.0)

Einstein wouldn't change his mind about cosmic stability until 1931—after he learned what the astronomers had been up to in the meantime, and realized that all the observational evidence was against him.

2.2　The Puzzle of the Nebulae

The observational path to the expanding universe had begun with perhaps the most vexing astronomical mystery of the late 19th and early twentieth centuries: spiral nebulae.

These nebulae were faint, fuzzy pinwheels of light found by the thousands all over the sky. Although one or two of them could just barely be seen with the naked eye—the most famous being a gossamer patch of light in the constellation of Andromeda (Fig. 2.4)—the vast majority were visible only through a telescope, and had no names beyond catalog numbers such as NGC 151.

But what were they? The argument raged for decades, with astronomers divided along lines laid out most famously on April 26, 1920, during a "Great Debate" at the then-new Smithsonian Museum of Natural History in Washington, DC [20]. Advocating for what was probably the majority view,

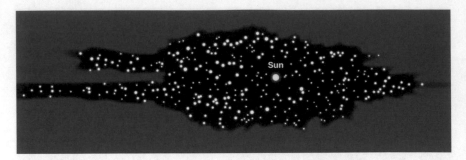

Fig. 2.4 The first hint of our stellar neighborhood's true structure came in 1785, when the astronomer William Herschel counted the number of stars he could see in each direction through his telescope and concluded that they formed a flattened disk aligned with the band of light we call the Milky Way. As can be seen in this diagram adapted from his paper, Herschel assumed that our Sun is near the center. (Franknoi A, Morrison D, Wolff SC (2016) Astronomy, Houston, Texas. Access for free at https://openstax.org/books/astronomy/pages/1-introduction, CC BY 4.0)

Harlow Shapley of the Mount Wilson Observatory in California held that the spirals were comparatively nearby objects—abnormal stars that had somehow become shrouded within whirlpools of gas, perhaps, or maybe brand-new solar systems caught in the act of forming. Heber Curtis of the Lick Observatory, also in California, argued for a kind of mega-Copernican view: just as the planets were other worlds and the stars were other suns, spiral nebulae were other "island universes" much like our own stellar neighborhood.

Bolstering the latter notion were observations dating as far back as 1785, when the German-English astronomer William Herschel decided to count the number of stars he could see in each direction with his telescope (Fig. 2.4). His conclusion was that formed a vast flattened disk aligned with the pale river of light known as the Milky Way [21]. Indeed, that's what the Milky Way *was*: an edge-on view of the disk as seen edge-on from the perspective of the Solar System. And of course, that was exactly what you'd expect any alien observers to see when they looked out at their own island universe.

For Shapley, however, the most compelling counterargument to this notion was its sheer implausibility.[4] For these other island universes to look so small and faint as seen from Earth, he pointed out, they would have to be millions of light years away—a distance that seemed ridiculously large.

---

[4] Another argument that seemed potent at the time was that at least one spiral nebula had been observed to rotate by a measurable amount within just a few decades—a fact that would have implied rotation velocities faster than the speed of light if the nebula were truly as far away as the island-universe advocates claimed. Only later was the observation found to be incorrect.

2.2.1 Celestial Fingerprints

The Great Debate was a draw, of course, since this was not politics but science, where disputes were supposed to be resolved with data. Fortunately, astronomers were already making rapid progress on that score, thanks to two key tools that were transforming their ability to measure what distant objects are made of, how far away they are from Earth, and how fast they are moving.

Fingerprints of Light

The first and oldest of these tools, *spectroscopy*, was based on the 19th-century discovery that atoms and molecules will emit or absorb light only at certain characteristic wavelengths.

These wavelengths are generally referred to as "lines" because that's what they looked like in 1814, when the Bavarian physicist and lens-maker Joseph von Fraunhofer first peered at the Sun with a new instrument he called the spectroscope. Through the eyepiece of this device—a prism-and-telescope combination that he had invented to study the colors of light in detail— Fraunhofer saw a thin sliver of sunlight spread sideways into a rainbow. But it was a rainbow interrupted by hundreds of vertical black lines, each representing a wavelength where the light was missing. He detected similar lines in the light from prominent stars, as well as *bright* lines in candle flames and the like.

Fraunhofer wasn't able to pursue this discovery very far, as his life was tragically cut short by toxic fumes from his glassmaking; he died in 1826 at the age of 39. But other scientists continued the work. And by the early 1860s, when the German scientists Gustav Kirchhoff and Robert Bunsen were pioneering the use of spectroscopy for chemical analysis, it was clear that the bright and dark lines produced by any given substance were like photographic negatives of one another. If you heated a sample of some atom or molecule in a flame— easily done with a gas-fueled "burner" devised by Bunsen—it would emit a pattern of bright lines. But if you placed a gaseous form of the sample in front of a brighter source, the gas would absorb light at those same wavelengths and form an identical pattern of dark lines. Indeed, this pattern could identify the substance as reliably as a fingerprint [22, 23] (Fig. 2.5).

An Astonishingly Rich Source of Information

The full explanation for these lines wouldn't be available until the 1920s, when physicists finally developed a comprehensive quantum theory of atomic structure. But nineteenth century astronomers didn't need to wait for the *why*

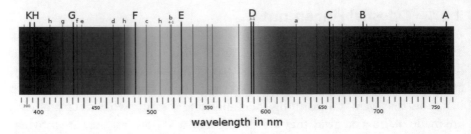

Fig. 2.5 The major lines that Joseph van Fraunhofer first observed in the spectrum of the Sun, labeled with their modern names. The 'nm' unit for their precise wavelengths stands for "nanometer," or one one-billionth of a meter (Image credit: public domain)

of the lines. Observers quickly realized that, whatever the lines were, they could be an astonishing rich source of information about sources that would otherwise be utterly out of reach.

What are stars and nebulae made of? By the end of the 1860s, a comparison between the lines seen in in these celestial object and emission lines seen in the laboratory had revealed that the heavens were made of exactly the same chemical elements found on Earth.[5] True, the proportions were often different—but it was the same stuff, nonetheless.

Do the stars spin like the Sun and planets do? The answer was clear as soon as astronomers looked at fine details like the width of the spectral lines: Yes, the stars do spin—some of them very fast.

And are the stars and nebulae as eternally fixed as they seem? Here the answer was equally clear: No. At least when it came to an object's motion toward or away from us, astronomers could get an accurate measure of its velocity simply by looking at how far the lines were shifted toward the blue or red end of the spectrum.[6] The first attempt to use this method on a star, published in 1868, found that the bright star Sirius was receding from us at an estimated 47 km per second [24]. And that was not at all unusual, as it turned out: Later observations showed that most of the stars we see in the night sky are buzzing around randomly at speeds at least that high—roughly 100 times faster than a rifle bullet.

---

[5] Observations made during a solar eclipse in 1868 were an exception that proved the rule. When observers saw a series of spectral lines on the Sun that they couldn't identify from the lab, they concluded that the lines belonged to a previously unknown element. They called it helium, after the Greek word for Sun, *helios*. It was first detected on Earth in 1881, in fumes from Mount Vesuvius.

[6] Side-to-side motions are much tougher to measure. Even the closest stars are so far away that it will typically take one of them roughly 2000 years to change its position relative to the more distant stars by the width of the full Moon.

Well into the twentieth century, unfortunately, spiral nebulae continued to be tricky subjects for this kind of analysis: They were so faint and diffuse that it was hard to get a spectrum at all, much less see lines. But in 1912, using a state-of-the-art instrument at the Lowell Observatory in Arizona, staff astronomer (and later, director) Vesto Slipher was finally able to get good readings on the Andromeda nebula—and was astonished to find its lines strongly blue-shifted, meaning that this spiral was moving toward us at a speed he estimated at 300 km per second[7]—so much faster than the motion of any known star that Slipher had to wonder if he'd made a mistake [25–27].

He hadn't; other astronomers soon confirmed his result. But as Slipher collected spectra from dozens of additional spiral nebulae, he did find that Andromeda's blueshift was the exception rather than the rule. Aside from a few other bright spirals, which are now known to be relatively nearby, every such nebula was *red* shifted—meaning that it was moving away from us, in some cases by more than 1000 km per second [28]. In 1914, Slipher also determined that the pinwheel appearance of the spirals was no accident: By measuring very tiny variations of the redshift in different parts of each spiral, he could see that one side was always a tiny bit redder than average, while the other side was a tiny bit bluer—which was exactly what you'd expect if the spiral structure was rotating, with one side moving away from us and the other side moving towards us. And more than that, Slipher found, the nebulae were always rotating with their spiral arms trailing behind, like a swirl of cream in a cup of coffee [29].

The cause of these large redshifts was an utter mystery at the time. But they did bring Slipher into the island-universe camp. He reasoned that the data simply didn't make sense unless the spirals were far, far away: any nearby object that was moving that fast would have escaped the Milky Way long ago, and wouldn't be nearby anymore.

2.2.2 A Cosmic Yardstick

Still, that argument by itself didn't *prove* anything. To nail down the island-universe hypothesis beyond a doubt, somebody would have to measure the spirals' actual distances. And distance measurement, happily, was the other astronomical tool that was progressing rapidly.

---

[7] The modern figure is 110 km per second.

Parallax and Its Limits

Until just a few years earlier, the only way to find celestial distances was with *parallax*, which is a geometric effect that's easy to see: just hold up a finger in front of your face, and then alternately close your right and left eyes. Notice how your finger seems to jump back and forth relative to objects on the far side of the room, by an amount that increases as you move your finger in, and decreases as you move it out. Much the same thing would happen if you made simultaneous observations of the moon from the opposite sides of the Earth: the two views would show the moon shifted relative to the distant stars. From there, it would just be a matter of elementary geometry to calculate the moon's distance, which is about 234,000 km. Indeed, Ancient Greek astronomers had used a parallax measure based on lunar and solar eclipses to get a pretty accurate estimate of the moon's distance more than 2000 years ago [30].

By the eighteenth century, observers had used variations on the parallax idea to work out most of the distances within the solar system. And they kept trying to do the same thing with stars, since the nearest would presumably show a parallax shift relative to the more distant ones. But stars proved to be a much tougher proposition: even the closest among them were much, much further away than the moon; otherwise, astronomers would see them wiggle from side to side over the course of a single night, as the Earth's rotation carried their telescopes from one side of the globe to the other. Indeed, the stars seemed to stay resolutely fixed in place even when astronomers made their observations six months apart, allowing the Earth to move to the other extreme of its orbit around the Sun and separate the measurements by the full 300-million-kilometer diameter of that orbit.

All of which is why the first compelling evidence for stellar parallaxes had to wait for the advent of bigger and better instruments in the nineteenth century [30]: even the largest shifts amount to no more than a fraction of an arcsecond—an angular measure that's equal to just 1/3600 of a degree. (In modern terms, that's about the width of a US quarter viewed from 5 km away.) Or to put it another way, even the closest stars are further away than one *parsec*, or 3.26 light years: the distance that will yield an annual parallax of exactly one arcsecond.

Credit for the discovery usually goes to the German astronomer Friedrich Bessel, who in 1838 published observations of the faint star 61 Cygni showing an annual parallax of 0.31 arcseconds. (That's close to the modern value, which corresponds to a distance of 11.4 light years or 3.498 parsecs.) Progress was slow, however; astronomers didn't begin to accumulate stellar parallax measurements in large numbers until the early twentieth century, after they had mastered the art of putting cameras on their telescopes and could begin to make very accurate measurements of stellar positions on glass photographic plates. And even then, the inevitable uncertainties in the measurements made parallax essentially useless for determining distances beyond a few hundred light years—not nearly far enough to settle the spiral nebula question.

A Standard Candle

What finally began to smash that distance barrier was a 25-page paper published in 1908 by Henrietta Swan Leavitt, one of a group of women staffers hired by the Harvard College Observatory to perform supposedly mundane tasks such as measuring and cataloging objects on the university's rapidly expanding collection of photographic plates [31]. For this particular article, Leavitt had compiled all the known data on 1,777 variable stars—the kind that brighten and dim on a more or less set schedule (Figs. 2.6).

There are many different kinds of variable stars, they are found all over the sky, and there was nothing unique about the ones in Leavitt's survey—except that they were all located in the Large and Small Magellanic Clouds, irregular swarms of stars that shine in the southern sky like detached pieces of the Milky Way [32]. Most of these variables were new to science, having been discovered since 1904 in plates taken at Harvard's observing station in Peru. But with so much fresh data in hand, Leavitt noticed a striking regularity: "It is worthy of notice," she wrote, "that...the brighter variables have the longer periods."

Fig. 2.6 Henrietta Swan Leavitt meticulously compiled all the available data on thousands of variable stars, and discovered the key to measuring the universe (Image credit: Harvard College Observatory, CC0)

Leavitt chose not to draw any hard conclusions from this fact, given that the available brightness estimates were still crude and preliminary. But she followed up in 1912 [33] with a detailed study of 25 carefully selected variables in the Small Magellanic Cloud, a relatively compact object whose stars were all at nearly the same (albeit unknown) distance from Earth.[8] These variables were of a type known as Cepheids, which wax and wane over the course of days or weeks. And Leavitt found that the relation between each variable's period and its average brightness was very clear and consistent: measure either quantity, and you would know the other.[9]

Furthermore, Leavitt wrote, "since the variables are probably at nearly the same distance from the Earth, their periods are apparently associated with their actual emissions of light"—a quantity also known as the star's intrinsic brightness, or *luminosity* (Fig. 2.7).

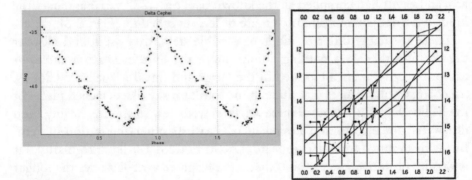

Fig. 2.7 The rising and falling brightness of Delta-Cephei, the star that gave its name to Cepheid variables (Image credit: ThomasK Vbg/Wikimedia Commons, CC BY-SA 3.0) **right** A scatter diagram from Leavitt's 1912 paper, in which she plots 25 Cepheids in the Small Magellanic Cloud at their brightest (upper line) and at their dimmest (lower line). The numbers on the vertical axis refer to each star's apparent magnitude, a measure of its brightness as seen from Earth; larger values mean dimmer stars, and vice versa. The numbers on the horizontal axis are the logarithm of each Cepheid's period of oscillation. Leavitt's discovery of this tight relation between period and luminosity turned Cepheids into the most important "standard candles" in astronomy (Image credit: Harvard College Observatory, Public Domain.)

---

[8] Although the paper was submitted on Leavitt's behalf by Harvard College Observatory director Edward Pickering, and appears over his name alone, its first sentence is quite clear about the real author's identity: "The following statement about the periods of 25 variable stars in the Small Magellanic Cloud has been prepared by Miss Leavitt".

[9] To a good first approximation, Leavitt found, a Cepheid's magnitude is proportional to the logarithm of its period.

This careful, measured statement had titanic implications: Cepheids could serve as "standard candles" for gauging distance. Anywhere in the sky that you found one, you just had to time its dimming and brightening to measure its period, then use Leavitt's relation to find its true luminosity. From there, finding the Cepheid's distance would be a simple matter of comparing that *true* luminosity to its *apparent* luminosity—that is, how bright or dim the star looked through the telescope.[10]

Or rather, this comparison *would* have been simple if anyone could determine the true luminosity of Leavitt's variable stars, which she herself couldn't do because the actual distance to the Small Magellanic Cloud was unknown. But in her 1912 paper Leavitt expressed her hope that astronomers would fill that gap by measuring the distance to Cepheids that were close enough to use parallax. Observers rushed to do just that [34, 35]—among them Shapley. And by 1918, two years before he participated in the Great Debate, Shapley had used Leavitt's relation to make an astonishing discovery about the Milky Way star system [35, 36].

It was *big*.

Using what was then the world's largest telescope, a 100-inch (2.5 m) reflector that had been placed into operation just a few years earlier at the Mount Wilson Observatory above Pasadena, California, Shapley had started by identifying Cepheids in the photogenic balls of stars known as globular clusters. Then, using these stars as standard candles to find their true distance, and thus the clusters' true location in space, Shapley found that the Milky Way, the clusters, and the Magellanic Clouds together encompass a volume some 100,000 times larger than anyone had suspected. Furthermore, Shapley found that the Milky Way was indeed a vast, flattened disk—but that our solar system is most definitely not in the middle. Shapley's best guess put us roughly 60,000 light years out from the center, which lay in the direction of the constellation Sagittarius. (The modern figure is about 26,500 light years.)

In his papers, Shapley referred to this Brobdingnagian structure as the *galaxy*, an old word for the Milky Way as seen with the naked eye. (It comes from the Greek term for "milky," *galaxias*.) But, as befit a leading member of the anti-island universe camp—which was an understandable position, given the mind-numbing scale of what he had found—Shapley described our galaxy as if it encompassed everything in the cosmos (Fig. 2.8).

It didn't. Half a decade after the Great Debate, at a meeting of the American Astronomical Society in 1925, a 34-year-old astronomer named Edwin Hubble detailed his observations of the Andromeda nebula and several other

---

[10] Any star's apparent brightness will follow an inverse square law: when it's twice as far away it will look one fourth as bright, and so on.

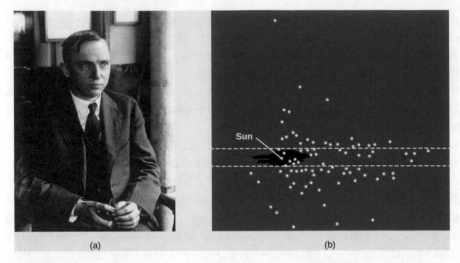

Fig. 2.8 a Harlow Shapley poses for a formal portrait. b In 1918, Shapley used the period-luminosity relation for Cepheid variables to map the true position of our galaxy's globular clusters, and found that they formed a rough sphere centered at a point in the constellation Sagittarius tens of thousands of light years away from our solar system. He concluded that this point is the true center of the galaxy, and that our Sun lies far from the center. The dark patch shows the approximate location of Herschel's diagram from 1785 (Franknoi A, Morrison D, Wolff SC (2016) Astronomy, Houston, Texas. Access for free at https://openstax.org/books/astronomy/pages/1-int roduction, CC BY 4.0)

spirals using the Mount Wilson 100-inch [37]. When photographed through this telescope, Hubble told his listeners, Andromeda's spiral arms turned out to be made not of gas, but of stars. And among those stars were Cepheids, which allowed him to apply Leavitt's period-luminosity relation.

Andromeda, said Hubble, lay some 930,000 light years from Earth—far outside even the huge Milky Way structure found by Shapley. (This was actually an underestimate. Astronomers would drastically revise the period-luminosity relation in the 1940s after they realized that there was more than one kind of Cepheid, leading to the modern figure of 2.5 million light years.)

Andromeda was an island universe in its own right—a whole other galaxy roughly as big as the Milky Way itself. And by extension, so were all the other spiral nebulae. Our universe, it was turning out to be vast beyond anything ever imagined.

2.2.3   A Universe That's Big and Getting Bigger

Hubble's discovery ended the great debate about spiral nebulae quite decisively—although it took a while for some island-universe skeptics to admit that. (Shapley did so almost immediately.) Hubble would continue to study galaxies until his death in 1953—showing, among other things, that the universe was also full of elliptical and irregular galaxies that had no spiral arms whatsoever, just millions or billions of stars.

That came later, however. In the 1920s, Hubble continued to gather distance data on additional spirals—often in collaboration with Mount Wilson staffer Milton Humason [30]. And somewhere along the way, he took galaxies for which he had decent distance data and plotted those values versus the mysterious redshifts measured a decade earlier by Slipher.

The plot had a lot of scatter, but showed a clear linear trend: the further away a galaxy was, the faster it was receding from us [38] (Fig. 2.9).

Considering that this result is now hailed as one of the most transformational discoveries in cosmology—and is a big reason why Hubble is widely considered to be the greatest astronomer of the twentieth century—it's remarkable how little attention his announcement got at the time [26]. Almost no one, including Hubble, knew quite what to make of it.

One the few who did was Georges Lemaître (Fig. 2.10), a Belgian physics professor who was a rarity: not only was he a Catholic priest, but he was one

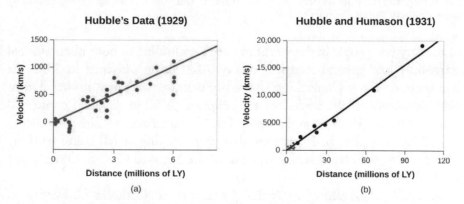

Fig. 2.9 a Adapted from the data in Hubble's 1929 paper in the Proceedings of the National Academy of Sciences. The vertical axis expresses each galaxy's redshift in terms of its velocity away from Earth. There is lots of scatter in the data, but the linear trend is evident. **b** Adapted from the data in Hubble and Humason's 1931 article in the Astrophysical Journal. The red dots in the lower left are the galaxies from Hubble's 1929 survey. The linear distance-velocity relation clearly extends to vaster greater distances. (Franknoi A, Morrison D, Wolff SC (2016) Astronomy, Houston, Texas. Access for free at https://openstax.org/books/astronomy/pages/1-int roduction, CC BY 4.0)

Fig. 2.10 Georges Lemaitre with Albert Einstein during a 1933 visit to Caltech. On the left is Caltech president Robert Millikan, winner of the 1923 Nobel Prize for measuring the charge of the electron. (Credit: unknown photographer, probably from Caltech, Public Domain)

of the very few people in the world to have a grounding in both observational astronomy *and* general relativity theory. After being ordained in 1923, he had spent a year at Cambridge University working with astronomer Arthur Eddington, one of the leaders of the eclipse expedition that had confirmed general relativity in 1919, and author of the first textbook on Einstein's theory [39]. Then Lemaître had spent another year studying at MIT and working with Shapley, who had become director of the Harvard College Observatory in 1921 [40].[11]

In 1927, after taking a teaching post at the Catholic University of Louvain, Lemaître had expanded upon Einstein's general-relativistic model of cosmology. And he had independently reached much the same conclusion that Friedmann had in 1924, when the Russian scientist had calculated that

---

[11] One of Shapley's first acts as director had been to name Henrietta Leavitt as head of stellar photometry at the Observatory. Sadly, she was already quite ill, and died of stomach cancer in December 2021.

the universe is almost certainly not static. Lemaître's equations showed that, unless the universe is very precisely balanced, a la Einstein, it will either be expanding or contracting—with or without a cosmological constant.

Unlike Friedmann, however, Lemaître had pointed out a striking observational consequence of that fact: if the universe is expanding, then the cosmic sphere will get bigger and bigger over time and take the galaxies along for the ride. So from the vantage point of any one galaxy—for example, the Milky Way—every other galaxy will appear to be receding with a velocity proportional to its distance.[12] Lemaître even used Slipher's redshift measurements to estimate the constant of proportionality. Thus the title of his paper, which in English reads, "A Homogeneous Universe of Constant Mass and Growing Radius Accounting for the Radial Velocity of Extragalactic Nebulae" [41].

Unfortunately, Lemaître had written his paper in French, and published it in a small Belgian journal that few astronomers saw. And he didn't do much to promote it, because Einstein had been quite discouraging. When Lemaître told the great physicist about his work at a conference later in 1927, Einstein praised the younger man's mathematics—but went on to say, "from the physical point of view, that [result] appeared completely abominable" [42].

Lemaître's paper didn't come to wider attention until January 1930, after he read about the confusion that Hubble's findings had caused at a meeting of the Royal Astronomical Society and sent a copy of the article to his former advisor Eddington at Cambridge. The English astronomer immediately recognized what Lemaître had accomplished and became the young priest's ardent champion. He arranged to have the 1927 article translated into English [43]. And in May 1930, Eddington published a widely read commentary of his own in which he not only raved about Lemaître's results, but pointed out that the equations had an additional consequence: Einstein's static-universe model of 1917, the one with the cosmological constant, was about as stable as a pencil balanced on its point. The slightest perturbation would send it toppling one way or another, into expansion or contraction [44]. And, given that Slipher's redshift measurements showed that other distant galaxies are moving away from us, expansion it was.

Indeed, Hubble and Humason soon extended their observations out to galaxies lying many times further away than the 18 in their original sample, and found that the straight-line relation between distance and redshift continued unabated [45]. Writing v for the speed at which a galaxy is moving

---

[12] The classic metaphor is to imagine the galaxies as dots painted on the surface of a balloon that's being blown up. Pick a dot, and it will remain stationary relative to the patch of balloon surface that surrounds it—but it will nonetheless see the more distant dots getting further and further away.

away and D for its distance from Earth, and using the modern notation H_0 for the constant of proportionality, this straight-line relation reads:

$$v = H_0 D \qquad (2.2)$$

In time, and for reasons that are not entirely clear, Hubble would get sole credit for this discovery. The equation would come to be called "Hubble's Law," just as the constant H_0 would become "the Hubble parameter"— although Lemaître, and perhaps Slipher, should arguably get equal billing [26].[13] What is clear is that the relation continues to be of fundamental importance to cosmology—not least because measuring H_0 automatically yields an estimate for the age of the universe. The details vary from one cosmological model to the next, but the answer is always proportional to the inverse of H_0: the bigger the parameter is, the younger the universe is, and vice versa.

This was a bit awkward for Hubble and Humason in 1931: their value for H_0—558 km per second per million parsecs—would require the universe to be only 1 or 2 billion years old, which was tough to reconcile with the apparent age of the Earth and Sun. Only later would the previously mentioned recalibration of the Cepheid period-luminosity relation move the estimates for H_0 down into the modern range of roughly 70 km per second per million parsecs, corresponding to a cosmic age estimate of 13.8 billion years.[14]

But again, all that would come later. In the meantime, Einstein's belief in a static universe was crumbling. It's hard to say which loomed larger in his mind—Eddington's finding that his 1917 solution was unstable, or a visit Einstein paid to Hubble and his colleagues in Pasadena in early 1931, when he learned about the evidence for an expanding universe firsthand. But in March of that year, Einstein wrote in a report to the Berlin Academy of Sciences that he had been wrong, that the work of Hubble had changed everything, and that "the assumption of a static nature of space is no longer justified" [16].

The universe was indeed expanding.

---

[13] The International Astronomical Union voted in 2018 to make the change official: the relation is now known as the "Hubble-Lemaître law.".

[14] As we'll see in Chap. 6, though, getting the value of H_0 any more precise than that continues to be a matter of intense controversy, calling into question our estimates of the age of the universe, and perhaps suggesting we need to rethink our models of the early universe.

References

1. Mahon B (2004) The man who changed everything: the life of James Clerk Maxwell. Wiley
2. Maxwell JC (1861) On physical lines of force. Philos Mag 21 & 23 Series 4:
3. Hertz H (1887) Ueber einen Einfluss des ultravioletten Lichtes auf die electrische Entladung. Ann Phys 267:983–1000. https://doi.org/10.1002/andp.188 72670827
4. Hertz H (1888) Ueber die Einwirkung einer geradlinigen electrischen Schwingung auf eine benachbarte Strombahn. Ann Phys 270:155–170. https://doi.org/10.1002/andp.18882700510
5. Hertz H (1888) Ueber die Ausbreitungsgeschwindigkeit der electrodynamischen Wirkungen. Ann Phys 270:551–569. https://doi.org/10.1002/andp.188 82700708
6. Brand JCD (1995) Lines of light. CRC Press
7. Einstein A (1905) Zur Elektrodynamik bewegter Körper. Ann Phys 322:891–921. https://doi.org/10.1002/andp.19053221004
8. Einstein A (1989) The collected papers of Albert Einstein, Volume 2 (English)
9. Einstein A (1905) Ist die Trägheit eines Körpers von seinem Energieinhalt abhängig? Ann Phys 323:639–641
10. Einstein A (1997) The collected papers of Albert Einstein, Volume 6 (English)
11. Einstein A (1915) Zur allgemeinen Relativitätstheorie. Sitzungsberichte K Preußischen Akad Wiss Berl 778–786
12. Einstein A (1916) Näherungsweise Integration der Feldgleichungen der Gravitation. Sitzungsber Preuss Akad Wiss Berl Math Phys 688–696
13. Catchpole R, Dolan G (2020) General relativity and the 1919 solar eclipse. In: Gen. Relativ. 1919 Sol. Eclipse. http://www.royalobservatorygreenwich.org/art icles.php?article=1283
14. Dyson FW, Eddington AS, Davidson C (1920) A determination of the deflection of light by the sun's gravitational field, from observations made at the total eclipse of May 29, 1919. Philos Trans R Soc Lond Ser Contain Pap Math Phys Character 220:291–333. https://doi.org/10.1098/rsta.1920.0009
15. Einstein A (1917) Cosmological Considerations in the General Theory of Relativity. Sitzungsber Preuss Akad Wiss Berl Math Phys 1917:142–152
16. Nussbaumer H (2014) Einstein's conversion from his static to an expanding universe. Eur Phys J H 39:37–62. https://doi.org/10.1140/epjh/e2013-40037-6
17. Friedman A (1922) On the curvature of space. Z Phys 10:377–386. https://doi.org/10.1007/BF01332580
18. Friedmann A (1924) Über die Möglichkeit einer Welt mit konstanter negativer Krümmung des Raumes. Z Für Phys 21:326–332
19. Friedmann A (1999) On the possibility of a world with constant negative curvature of space. Gen Relat Grav 31:2001–2008

20. Shapley H, Curtis HD (1921) The scale of the universe. Bull Natl Res Counc 2:171–217
21. Herschel W (1785) On the construction of the heavens. By William Herschel, Esq. F. R. S. Philos Trans R Soc Lond 75:213–266
22. Kirchhoff G, Bunsen R (1860) Chemische Analyse durch Spectralbeobachtungen. Ann Phys 186:161–189. https://doi.org/10.1002/andp.18601860602
23. Kirchhoff G, Bunsen R (1861) Chemische Analyse durch Spectralbeobachtungen. Ann Phys 189:337–381. https://doi.org/10.1002/andp.18611890702
24. Huggins W (1868) Further Observations on the Spectra of Some of the Stars and Nebulae, with an Attempt to Determine Therefrom Whether These Bodies are Moving towards or from the Earth, Also Observations on the Spectra of the Sun and of Comet II., 1868. Philos Trans R Soc Lond Ser I 158:529–564
25. Slipher VM (1913) The radial velocity of the Andromeda Nebula. Lowell Obs Bull 2:56–57
26. O'Raifeartaigh C (2013) The Contribution of V. M. Slipher to the discovery of the expanding universe. In: Origins of the expanding universe: 1912–1932. Astronomical Society of the Pacific, San Francisco, p 49
27. Nussbaumer H (2013) Slipher's Redshifts as Support for de Sitter's Model and the Discovery of the Dynamic Universe. In: Origins of the Expanding Universe: 1912–1932. Astronomical Society of the Pacific, San Francisco, p 25
28. Slipher VM (1917) Nebulae. Proc Am Philos Soc 56:403–409
29. Slipher VM (1914) The detection of nebular rotation. Lowell Obs Bull 2:66–66
30. Webb S (1999) Measuring the Universe: the cosmological distance ladder, 1999 edition. Springer, London ; New York : Chichester, UK
31. Sobel D (2017) The glass Universe: how the ladies of the Harvard Observatory Took the Measure of the Stars, Reprint edition. Penguin Books
32. Leavitt HS (1908) 1777 variables in the magellanic clouds. Ann Harv Coll Obs 60(87–108):3
33. Leavitt HS, Pickering EC (1912) Periods of 25 variable stars in the small magellanic cloud. Harv Coll Obs Circ 173:1–3
34. Hertzsprung E (1913) Über die räumliche Verteilung der Veränderlichen vom δ Cephei-Typus. Astron Nachrichten 196:201
35. Shapley H (1918) Studies based on the colors and magnitudes in stellar clusters. VII. The distances, distribution in space, and dimensions of 69 globular clusters. Astrophys J 48. https://doi.org/10.1086/142423
36. Shapley H (1918) Globular clusters and the structure of the galactic system. Publ Astron Soc Pac 30:42. https://doi.org/10.1086/122686
37. Hubble EP (1925) Cepheids in spiral nebulae. Observatory 48:139–142
38. Hubble E (1929) A relation between distance and radial velocity among Extra-Galactic Nebulae. Proc Natl Acad Sci 15:168–173. https://doi.org/10.1073/pnas.15.3.168
39. Eddington AS (1923) The mathematical theory of relativity. University Press, Cambridge

40. (1966) Georges Lemaitre. Phys Today 19:119. https://doi.org/10.1063/1.304 8455

41. Lemaître G (1927) Un Univers homogène de masse constante et de rayon crois-sant rendant compte de la vitesse radiale des nébuleuses extra-galactiques. Ann Société Sci Brux 47:49–59

42. Luminet J-P (2013) Editorial note to: Georges Lemaître, A homogeneous universe of constant mass and increasing radius accounting for the radial velocity of extra-galactic nebulae. Gen Relativ Gravit 45:1619–1633. https://doi.org/10.1007/s10714-013-1547-4

43. Lemaître G (1931) Expansion of the universe, A homogeneous universe of constant mass and increasing radius accounting for the radial velocity of extra-galactic nebulae. Mon Not R Astron Soc 91:483–490. https://doi.org/10.1093/mnras/91.5.483

44. Eddington AS (1930) On the instability of Einstein's spherical world. Mon Not R Astron Soc 90:668–678. https://doi.org/10.1093/mnras/90.7.668

45. Hubble E, Humason ML (1931) The velocity-distance relation among Extra-Galactic Nebulae. Astrophys J 74:43. https://doi.org/10.1086/143323

3

The Discovery of the Big Bang

Even after the expansion of the universe was discovered in the 1930s, the notion that it implied a cosmic beginning was far from clear. For most astronomers of the day, the question of where the universe came from was a non-issue—a fine topic for philosophers, theologians, and late-night dorm debates, maybe, but much too nebulous a subject for serious researchers [1].

Perhaps astronomers shared Einstein's deep-seated belief in a static universe. Or perhaps they just didn't want to face the inevitable follow-up question: If the universe had a "beginning," then where did that beginning itself come from? This seemed to be a mystery that was unanswerable by any known physical law.[1]

Indeed, this reluctance to contemplate cosmic origins was so entrenched that the dramatic cosmic origin we now call the Big Bang had to be rediscovered three times before the idea finally took hold.

3.1 Bright but Very Rapid Fireworks

The first discovery was triggered in January 1931, when Arthur Eddington—the Cambridge astronomer who had found early observational evidence for Einstein's general relativity—gave a presidential address to a UK educational group called the Mathematical Association. In the course of his talk he

---

[1] We will return to this mystery in Chap. 4, where we see how scientists are attempting to address it.

© The Author(s), under exclusive license to Springer Nature Switzerland AG 2022
M. M. Waldrop, *Cosmic Origins*,
https://doi.org/10.1007/978-3-030-98214-0_3

admitted that yes, it was logically possible for time to have had a beginning. But philosophically, said Eddington, "the notion of a beginning of the present order of Nature is repugnant to me" [2].

In Louvain, the Belgian priest, physicist and astronomer Georges Lemaître saw the "repugnant" comment two months later, when the British journal *Nature* published Eddington's address in full, and was inspired to reply with a short letter to the editors of *Nature* [3].

He didn't find the notion of a beginning repugnant at all, Lemaître's letter declared.[2] If nothing else, he said, consider the newly revealed physics of the quantum realm, where small particles behave in ways alien to our everyday experience. Because of that weirdness, Lemaître wrote, "I would rather be inclined to think that the present state of quantum theory suggests a beginning of the world very different from the present order of Nature."

After all, Lemaître explained, as time passes the total energy of the universe must inevitably get subdivided among more and more "quanta"—what we'd now call elementary particles such as photons and electrons. Indeed, this is just one version of the second law of thermodynamics, which famously says that the universe as a whole becomes increasingly disordered over time. So if we imagine the clock running backwards toward the beginning, said Lemaître, "we must find fewer and fewer quanta, until we find all the energy of the universe packed in a few or even in a unique quantum"—an entity whose decay would give rise to the universe as we see it today.

And what came before the first quantum? The question was meaningless, Lemaître suggested. You can't ask about "before" unless time already exists. But in a quantum universe, he said, sounding very much like a 21st-century quantum-gravity theorist, it's likely that space, time, and matter came into being together, emerging as collective phenomena from the behavior of groups of quanta.

Lemaître's guess was that this first quantum was some kind of primordial atom: A titanic atomic nucleus that contained the entire mass of the universe, and that decayed via a kind of super-radioactivity to yield the cosmos we now live in.

This guess was wildly off-base—which was hardly surprising, given how little was known about nuclear physics in 1931. But being wrong about the specifics doesn't detract from Lemaître's larger accomplishment: In a letter to the editor totaling only a few hundred words, he had given the scientific literature its first recognizable description of what we now call the Big Bang. And more than that, simply by embracing the idea of a beginning that unfolded

---

[2] Lemaître's reaction had nothing to do with the Biblical creation account in *Genesis*; although he was a Catholic priest, he was always adamant about keeping his faith separate from his science.

according to natural law, Lemaître had become the first to assert that the problem of cosmic origins could be studied by the methods of science.

Lemaître elaborated on this idea just a few months later with the publication of a longer *Nature* essay, which included two more prescient suggestions [4]. The first was that the initial expansion of the universe must have been very fast, and that the matter it contained must have been very hot. When we look at our own galaxy and its nearest neighbors, he wrote, "they are ashes and smoke of bright but very rapid fireworks." This hypothesis would turn out to be correct: much of the later research in cosmology has been a quest to understand exactly what happened during that initial fireworks display.

Lemaître's second prescient suggestion was that the initial fireworks would produce a kind of afterglow: radiation that would still be streaming down from the sky. Here too, Lemaître was wrong on the specifics; he believed that this afterglow is the source of the celestial particles known as cosmic rays, which are now thought to originate much later in cosmic evolution. But again, his larger insight was correct: There *is* a Big-Bang afterglow—although, as discussed below, it wouldn't be detected for another three decades.

In the meantime, however, Lemaître's concept of cosmic origins sank into immediate obscurity, if only because few researchers wanted to believe in something as weird as a primordial atom without solid observational evidence [1].

Indeed, the second discovery of the Big Bang would have to wait for a decade and a half, while the rise of fascism devastated European science, civilization tried to tear itself apart in the Second World War—and a few scientists, steeped in the fast-moving field of nuclear physics, began to realize that we've been surrounded by that solid observational evidence all along.

3.2   The Fireball's Fossils

3.2.1   The Star-Stuff Conundrum

As with many scientific discoveries, the road to this one started with a seemingly unrelated question: what makes the Sun shine? What makes *any* star shine?

It had been obvious since the nineteenth century that neither the Sun nor any other star could be burning like a candle flame. Not only is there no oxygen in space to support combustion, but the Sun would have exhausted any available fuel and burned out long ago, making it hard to account for all those millions of years of fossils and geologic strata seen here on Earth. But if not combustion, then what?

In 1920, Arthur Eddington suggested a new possibility [5]. Whatever he may have thought about cosmic origins, Eddington had always been among the first to see how the latest advances in physics might apply to astronomy. In this case, he knew that physicists had spent the previous decade establishing that every atom in the universe had a similar structure, consisting of a tiny, dense, positively charged *nucleus* surrounded by a fluffy cloud of negatively charged electrons. He also knew that the lightest element, hydrogen, had a nucleus that was almost exactly one quarter of the mass of the second lightest element, helium.

Almost, that is—but not quite. According to the latest high-precision laboratory measurements, Eddington noted, the mass of four hydrogen nuclei taken together was actually a bit larger than the mass of one helium nucleus. This tiny discrepancy led him to wonder—what if the hot, dense conditions inside the Sun somehow allowed those four hydrogens to fuse into a single, slightly less massive helium atom? The excess mass would have to go somewhere, Eddington reasoned. And indeed, the answer was right there in Einstein's celebrated equation $E = mc^2$: the tiny missing mass would be converted into a huge burst of energy. (Remember that c stands for the speed of light, which is a very large number—and its square is larger still.)

Assuming that this is what's happening, Eddington wrote, "we need look no further" for the power source of the stars.

Eddington's inspired guess would eventually prove correct: Today this process is called thermonuclear fusion, and it does indeed power the stars. But proving this would take a while. Not only were the details of such nuclear reactions hopelessly fuzzy in 1920, but astronomers of the day were convinced that neither the Sun nor any other star contained nearly enough hydrogen to keep the lights on for billions of years. In fact, as best observers could tell from looking at stellar spectral lines,[3] most stars had a composition similar to Earth's, just hotter. And that meant lots of heavy elements like iron, with only a trace of light elements like hydrogen.[4]

That interpretation wouldn't last much longer, however, thanks to Cecilia Payne (Fig. 3.1), a British student who had become fascinated with physics and astronomy after hearing one of Eddington's lectures at Cambridge

---

[3] As discussed in Chap. 2, these features in the light of a star (or the Sun, or any other celestial body) are also known as Fraunhofer lines, after their discoverer. In principle, the pattern of lines can reveal how much of each chemical element a star contains.

[4] There is plenty of hydrogen in the ocean, it's true. But from a planetary perspective, Earth's oceans are just a thin film of moisture on the surface of dense ball that's mostly rock and iron.

Fig. 3.1 In 1925 Cecilia Payne-Gaposchkin became the first woman to earn an astronomy Ph.D. in the United States. Her dissertation showed observers how to measure the true cosmic abundance of the chemical elements (Credit: Harvard College Observatory, CC0)

University.[5] Payne came to America in 1923, when director Harlow Shapley at the Harvard College Observatory offered her a fellowship to study there. Shapley was so impressed that he soon started encouraging her to pursue a doctorate in astronomy—an endeavor that would take perseverance on her part and some energetic advocacy on his, given the still-strong resistance to women in the field [6, 7]. But Payne persisted, and in 1925 became the first woman to earn an astronomy Ph.D. in the United States—with a dissertation that has been called "the most brilliant Ph.D. thesis ever written in astronomy." [8, 9].

To interpret a star's spectral lines correctly, Payne argued, you first had to understand how those lines are actually produced [10, 11]. And to do *that*,

---

[5] Eddington's lecture was in November 1919, and covered the just-announced results of the eclipse expeditions that verified the gravitational deflection of starlight by the Sun – one of the key predictions of Einstein's theory of general relativity. The 19-year-old Payne was so electrified by what she'd heard that she went back to her room and wrote out the lecture from memory, practically verbatim. Eddington became one of Payne's most enthusiastic supporters, and encouraged her new dream of becoming an astronomer. He also wrote her a letter of introduction to Shapley – who had just started a fellowship aimed at helping women study at his Observatory.

she showed, you had to incorporate everything that physical theory and laboratory data could tell you: How the temperature, pressure, and gas density varied in the outer layers of a star; how the star's radiation streamed through those layers; how likely it was for an atom of any given element to absorb a bit of that radiation and become "ionized," or have an electron stripped away; how long it took for that ionized atom to recapture a stray electron—on and on.

The analysis was immensely complex, especially without digital computers. But after 215 pages of exhaustive analysis, Payne found that all those earlier composition estimates had to be turned upside down. A correct reading of the spectral data showed that hydrogen is by far the *most* abundant element in any star, followed by helium.[6] The heavier elements that we're familiar with on Earth, such as silicon, carbon, and iron, are present only in minor amounts.

Or to put it another way, most stars have more than enough hydrogen to keep shining.

3.2.2 Back to the Nucleon Soup

Payne, who would change her name to Cecilia Payne-Gaposchkin after her marriage in 1934, would soon turn her attention to other astronomical problems. But her spectral-analysis methods quickly became part of the standard tool kit of astronomy—and in due course, produced a slew of abundance numbers that forced observers to confront a whole new set of mysteries. Why, for example, did the composition of every star turn out to be essentially the same, with hydrogen and helium in roughly a roughly 3 to 1 ratio by mass?[7] Why do the heavier elements form such a tiny fraction of the cosmic total— maybe 2%? And within that heavy fraction, why are elements such as carbon or oxygen so much more abundant than elements such as lithium or boron?

Fortunately, the 1930s were a good time to be asking such questions, since physicists were finally getting somewhere in their efforts to understand the atomic nucleus. One of their major achievements was discovering that each

---

[6] Payne was actually talked out of stating this explicitly in her final published draft, give the still-rampant skepticism about the prevalence of hydrogen and helium in the Sun and stars. But the result was implicit in her calculations and data, and was eventually recognized widely.

[7] This is the ratio when you look at how much of a star's *mass* is contributed by each element. (The modern figures are that 75% of a typical star's mass is hydrogen, 23% is helium, and about 2% is in the form of heavier elements.) But of course, each helium atom is 4 times as massive as hydrogen, and the other elements are heavier still. When you look at the cosmic ratio of *atoms*, the results are even more dramatic: only 2 or 3 out of every thousand atoms are heavy elements, around 78 are helium—and 920 are hydrogen.

atomic nucleus contains a mix of positively charged *protons* that determine which element it is—one proton for hydrogen, two for helium, six for carbon, and so on—plus a roughly equal, but variable number of electrically neutral *neutrons* that determine which isotope it is.[8] The six protons in carbon, say, can be joined by six, seven, or eight neutrons to make carbon-12 (the most common isotope), carbon-13 (rare), or carbon-14 (unstable and radioactive.)

Holding these particles together is a "strong" force that only the protons and neutrons can feel; it has no effect whatsoever on the much lighter, negatively charged electrons that orbit the nucleus. This force is far stronger than gravity or electromagnetism—and it has to be, to keep the electrostatic repulsion between the positively charged protons from tearing the nucleus apart. Yet it is extremely short-range: the protons and neutrons in the nucleus practically have to be touching for the strong nuclear force to have any effect at all.

Taken together, these properties explained how new elements can form from existing nuclei—a "nucleosynthesis" process that boils down to smashing things together and letting the protons and neutrons reshuffle themselves. Bare neutrons are particularly good at this. Being devoid of electric charge, they can sail right through an atom's protective electron cloud, plunge into the nucleus, and either merge to form a new isotope or knock pieces off to form an entirely new element.

But complex atomic nuclei can merge, too—not just hydrogen, as Eddington had guessed, but any number of elements. Getting such nuclei to merge is a much tougher proposition than with neutrons alone, though, mainly because of the electrostatic repulsion between the positively charged protons. The only way for nuclei to overcome this repulsion is to be moving so very, very fast that they can't be stopped in before they hit. Or equivalently, they have to be raised to temperatures measured in millions or even billions of degrees, which can happen only in extreme environments like the core of a star (Fig. 3.2).

---

[8] The existence of neutral particles in the nucleus had been hypothesized as far back as 1920, which is also when they got the name "neutrons." Individual neutrons were first observed in the laboratory in 1932, and as expected, proved to have a mass that was virtually identical to that of the proton. (The neutron's mass was slightly higher, actually.) Free neutrons also proved to be ever so slightly radioactive, decaying after about 14 min and 40 s on average into a proton, an electron, and a *neutrino*: a new, invisible particle that wouldn't be directly observed for decades yet. Neutrons inside certain atomic nuclei can fall apart like this, as well, producing a form of radioactivity known as *beta decay* – beta being an early name for radiation that consists of fast-moving electrons. But in most nuclei, the neutrons are bound so tightly that they are completely stable.

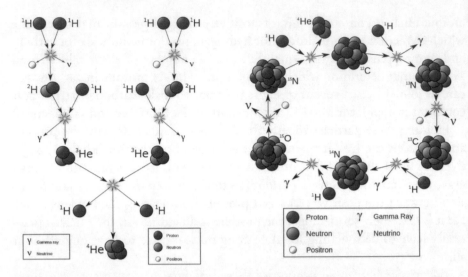

Fig. 3.2 As first suggested by Arthur Eddington in 1920, the fusion of four protons into a helium nucleus is the main energy source of comparatively low-mass stars like the Sun. In the 1930s, George Gamow, Hans Bethe and other realized that this is a complex, multistage process; the sequence shown here is just one of several possibilities. **right** In 1939, Bethe discovered a catalytic carbon–nitrogen-oxygen cycle that converts hydrogen into helium more efficiently, but requires higher temperatures. It is the main energy source for stars more massive than the Sun. (Credit: Borb, Wikimedia Commons, CC BY-SA 3.0)

On the other hand, as physicists also began to realize in the 1930s, this makes stars an obvious candidate for the birthplace of heavy elements—including those that make up the Earth and all of us who live on it. (Or, as Carl Sagan would famously put it decades later, "We are stardust.").

A leading advocate for this notion was George Gamow (Fig. 3.3), an irrepressible Russian physicist known for his wide range of interests and habit of tossing out ideas in a torrent—many of them ridiculous, but maybe 10% of them brilliant [12]. Gamow had made his reputation a decade earlier with some elegant theoretical work on radioactive decay [13]. But after 1934, when he defected from Joseph Stalin's oppressive Soviet Union and settled in the United States, he shifted his focus to nucleosynthesis in stars. And, being Gamow, he quickly attracted many other scientists to the problem—perhaps the most notable of whom was Hans Bethe, a German-emigre scientist whose insights into how stars produce energy would win him the 1967 Nobel Prize in physics [14].

Gamow's work on stars would slow down during the early 1940s, when he and most other physicists were caught up in the war effort. (Some worked on the Manhattan Project to develop an atomic bomb, while others went

Fig. 3.3 The Russian-born physicist George Gamow was the first to see how the cosmological equations of general relativity could be combined with modern nuclear physics. In 1948, he helped pioneer the study of what is now called the Big Bang. (Credit: American Institute of Physics)

to MIT to pursue the development of radar; Gamow himself consulted on explosives for the U.S. Navy.) But by 1946 he was back at it full time— and had started to shift his focus from stars to the early universe. Could this have been another birthplace for the chemical elements, he wondered?

In a sense, this question was just Gamow coming home to his roots: Back in Leningrad he had studied under the mathematician Alexander Friedmann, who was busy at the time developing the cosmological solutions to general relativity discussed in Chap. 2 [15, 16]. Back then, Friedman had shown that Einstein's equations were compatible with an expanding universe. Now, Gamow was in a position to combine Friedmann's general relativistic models of cosmology with cutting-edge nuclear physics.

After publishing an initial account of his ideas in 1946 [17], Gamow tackled the details in collaboration with two young colleagues, Ralph Alpher and Robert Hermann. And in an extraordinary burst of creativity, the three scientists in various combinations produced 11 publications on early-universe nucleosynthesis in 1948 alone [12, 18, 19]. (The first of the 1948 papers, written by Alpher and Gamow, is also famous for a non-physics reason: Gamow jokingly added Hans Bethe as the second author even though Bethe had not contributed to the work, so that when the list was read out loud—and when Bethe's name was given its correct German pronunciation, BAY-tuh—it would rhyme with the *alpha–beta-gamma* start of the Greek alphabet [20].)

In any case, by the end of 1948 Gamow, Alpher, and Hermann had arrived at a picture of cosmic origins that was nothing at all like Lematire's "unique quantum"—but was very close to modern thinking.

To see how they reached their conclusion, it helps to start with one of cosmologists' favorite tricks, and imagine the history of the universe running backwards from the present day toward the beginning. Think of watching a video in reverse, or listening to NASA count down toward a rocket launch.

For some number of billions of years this rewinding movie will actually be pretty boring, since nothing much is happening. True, the universe will be contracting rather than expanding, and the galaxies will be rushing together instead of separating. But they will they still be comfortably far apart.

But things will start to pick up once the countdown clock reaches a cosmic age of a billion years or so, which is the time-reversed era of galaxy formation. First, the in-rushing galaxies will start to crowd in on one another—and soon enough will begin to boil off and disassemble themselves into their original raw materials: mostly hydrogen and helium gas.

Then, as the shrinking universe carries that material further and further back toward the beginning, the gases will be compressed mercilessly. Denser and denser, hotter and hotter—the conditions will grow into something far beyond hellish. And keep going.

Finally, when the cosmic countdown gets to a few hundred thousand years from the beginning, the atoms in this cosmic gas will be jammed together so closely, and will be colliding so crazily, that their electrons will be stripped away from their nuclei and the very concept of an atom will lose its meaning. From here on in, all the matter in the universe will be a *plasma*, with electrons and nuclei moving independently. (Think of the glowing gases inside an old-style fluorescent tube.)

Yet the in-rushing universe will just be picking up speed. As the cosmic clocks relentlessly ticks down from days to hours to minutes and finally to seconds, the plasma's density will eventually come to exceed that of steel, or lead, or anything else humans have experienced. Likewise, the plasma's temperature will continue to skyrocket—and not just for matter; the contracting universe will compress and heat photons, the particles of light, as well. Finally, when the cosmic countdown finally gets to about 10 to 20 s, the photons will grow so dense and energetic that not even the strong nuclear force can resist them: they will shred the nuclei into a superheated soup of isolated protons or neutrons—particles often known generically as *nucleons*.

Gamow, Alpher, and Herman stopped their imaginary countdown at this point, since it was about as far as known physics could take them in 1948. And in any case, there seemed little point in pushing back any further: Friedman's cosmological equations showed that if they tried to extrapolate all the way down to $t = 0$, they would be faced with a universe compressed into a single point—a *singularity* of infinite density, infinite temperature, and zero size, where no known laws of physics could possibly apply. For much the same reason, the three physicists made no attempt to explain where the nucleon soup had come from, but simply took it as a given.

Still, extrapolating back to 10–20 s was a major advance in its own right. And a second finding followed, as Gamow, Alpher and Herman imagined running the clock forward again and traced how the initial nucleon soup would have evolved as the universe expanded. The calculations were horrendous—especially since they still didn't have electronic computers—and required a meticulous accounting of factors such as the rate of cooling, the rate at which neutrons decayed into protons and electrons, how radiation interacted with matter, and much more. In the end, however, Gamow, Alpher, and Herman arrived at a very simple bottom line: within just a few minutes after $t = 0$, the rapidly cooling brew would have condensed into nearly pure hydrogen and helium—with element- and isotope ratios close to those we see today [21].

And then there was a final finding. Alpher and Herman mentioned it almost in passing in late 1948, when they noted in a short letter to *Nature* that the radiation emitted from that primordial fireball would still be around, just cooled and redshifted by billions of years of cosmic expansion [22]. This was essentially what Lemaître had suggested in 1931, except that today we would see that radiation not as cosmic rays—Lemaître's guess—but as a faint wash of photons streaming down from every direction in the sky. Alpher and Herman were even able to estimate that the radiation's current temperature would be just a few degrees above absolute zero—a frigid value that would put the photons' wavelengths in the microwave region of the spectrum.

Although it's not clear whether Gamow, Alpher, and Herman actually knew of Lemaître's earlier work—they didn't cite it—it is fair to say that, in effect, they had taken the Belgian's intuitive speculations and made them rigorous with modern nuclear physics.

And yet, like Lemaître before them, Gamow, Alpher, and Herman also saw their work sink into obscurity.

Partly this was because they themselves were a bit disappointed: they had hoped to show that their primordial fireball generated all the elements, but found that, for reasons discussed below, the early universe couldn't produce much of anything beyond hydrogen, helium, and a tiny bit of element 3, lithium. Partly it was because there weren't many other people paying attention. Only a handful of astronomers and physicists worked on cosmology in those days—and even they tended to think of it as a hobby to dabble in when they weren't doing their *real* research.

Mostly, however, the three physicists' work fell into obscurity because the tiny community of part-time cosmologists was soon caught up in controversies over *another* cosmological theory from 1948—one that did away with the primordial fireball entirely.

3.2.3   Continuous Creation? or a Big Bang?

Sometime in 1946, as best they could later remember, three young Cambridge University physicists named Hermann Bondi, Thomas Gold, and Fred Hoyle found themselves kicking around ideas for an alternative approach to cosmology—one that would get rid of this notion of an evolutionary universe unfolding from some kind of cosmic beginning [23]. Like Eddington and many other scientists before them, the Cambridge trio shared a profound distaste for the very idea of a creation event, which seemed to be outside of natural law.

The problem, of course, was that if the universe really was expanding, as evidence seemed to suggest, then a cosmic beginning seemed unavoidable. So their bull sessions were going nowhere—until Gold thought of a resolution: What if empty space wasn't quite empty? What if the vacuum were somehow generating new matter—hydrogen atoms, perhaps—at some very low rate? If so, then as the universe grew and as galaxies moved apart, this newly generated matter would pop up out of empty space to fill in behind them and keep the average cosmic mass density constant. And the mutual gravitational attraction among these new atoms would eventually pull them together into new stars and galaxies to replace the old.

Like a river that is always flowing yet never moving, in other words, such universe would be always expanding yet never changing—and would not require an origin.

As Bondi later wrote, "Fred and I said, 'Ach, we will disprove this before dinner.' Dinner was a little late that night, and before long we all saw that this was a perfectly possible solution to this question." [24]. The obvious objection was that astronomers would surely have noticed if matter were popping into existence all around us. But a quick calculation showed that the continuous creation process would only need to generate a mass equivalent to a few new hydrogen atoms per cubic meter per million years—a rate that would be completely unobservable. [25] In fact, Bondi, Gold and Hoyle couldn't think of any astronomical observation that contradicted the assumption.

In 1948 they published two fleshed-out versions of this idea. Bondi and Gold called theirs a "steady-state" theory of the expanding universe, and based it on philosophical first principles [26]. Hoyle, who called his the "continuous creation" theory, simply modified the equations of general relativity with a term that allowed for matter production, and showed that it led to a static universe without the need for a cosmological constant [27].

Neither version attracted much notice at first. But that changed abruptly in March 1949, when the BBC invited Hoyle to discuss his ideas in a radio lecture aimed at the general public [23, 28]. Hoyle was already earning a reputation as a talented popularizer of science, in the manner later made famous by astronomers Carl Sagan and Neil DeGrasse Tyson. But his BBC broadcast is remembered today mainly for the way he described the evolutionary cosmologies he hated: "These theories were based on the hypothesis that all matter in the universe was created in one big bang at a particular time in the remote past," Hoyle said. A few minutes later he used the same phrase to disparage these cosmologies relative to his own model: "On scientific grounds this big bang hypothesis is much the less palatable of the two. For it is an irrational process that cannot be described in scientific terms." The following year, he used the phrase "big bang" pejoratively several more times in a series of BBC lectures that were later transcribed and published as a best-selling book, *The Nature of the Universe* [29].

Hoyle's breezy, alliterative shorthand didn't catch on right away; the phrase "Big Bang" wouldn't start appearing in scientific journals until the 1970s [30]. But his high-profile advocacy for the steady-state idea did ignite intense controversy among his fellow scientists. Some embraced the concept as an elegant way to get around the cosmic origin problem. Others despised it, on the grounds that continuous creation was an utterly ad-hoc assumption that violated some of the most fundamental principles of physics—not the least being the conservation of energy. If mass is created from nothing, after all, then $E = mc^2$ means that *energy* is created from nothing.

Either way, though, the steady-state controversy did force scientists to confront how little they actually knew about the universe at large, and to start looking for ways to fill the gaps. A particularly urgent question was to understand where the elements came from. In the steady-state model, the elements all had to be made from newly created atoms, be it hydrogen or whatever. But that required nucleosynthesis at extremely high temperatures. And since the steady-state model didn't allow for a primordial fireball, the only alternative site was in the stars. So for the steady-state idea to be viable, Hoyle and his fellow believers had to prove that this was possible—that nucleosynthesis in the stars could account for *all* the elements [31–33].

Frustratingly, however, nature didn't seem to be cooperating. As nearly as anyone could tell at the time, elements in the thermonuclear environment of a star would mostly grow by absorbing stray neutrons one by one, possibly followed by the emission of an electron to turn the captured neutron into a proton. This was fine for turning hydrogen into the lightest isotopes. The step-by-step capture would steadily build up deuterium, or "heavy hydrogen," an isotope with one proton and one neutron; helium-3, a rare form of that element containing two protons and one neutron; and helium-4, a very stable cluster of two protons and two neutrons that is also known as an alpha particle, and is by far the most abundant form.

But the process would slam into a brick wall after that: There is no stable isotope with a mass of 5. Even if such a nucleus got created by accident, it would fall apart within a fraction of a nanosecond—long before another neutron could wander in and turn it a heavier, more stable nucleus. Nor would it help to fuse two alpha particles into beryllium-8 (four protons and four neutrons), since that isotope falls apart even faster.

This barrier also applied to element creation in any kind of Big Bang scenario, which is why Gamow, Alpher, and Herman hadn't been able to account for any of the elements beyond helium. Still, those heavier elements obviously existed, so there had to be a way for nature to leapfrog that barrier. And in 1952, Cornell University physicist Edwin Salpeter showed how this happens [34]. Under the hot, dense conditions that prevailed at the core of certain red giant stars, he found, it was possible for three alpha particles to merge simultaneously (Fig. 3.4). The result would be a very stable carbon-12 nucleus, which has six protons and six neutrons. And from there, the way would be open for thermonuclear reactions in the stars to produce all the other heavy elements. Crucially, the necessary conditions would *not* occur during the Big Bang—and so the paper was widely viewed as indirect support for the steady-state model.

There was still a hitch, however. Salpeter's triple-alpha calculation predicted heavy-element ratios that didn't match the observed values, and

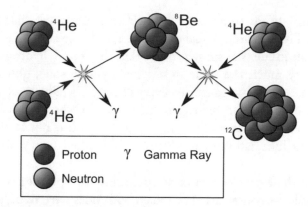

Fig. 3.4 The triple-alpha process discovered in 1952 leapfrogs the barriers at masses 5 and 8, where there are no stable isotopes. The process requires extremely high temperatures and densities found only in certain red giant stars. But it does allow such stars to make carbon-12—and from there, all the heavier elements beyond (Credit: Borb, Wikimedia Commons, CC BY-SA 3.0)

a rate of carbon formation that seemed way too low. But that discrepancy inspired Hoyle to make an audacious proposal: the ratios could be brought into line and carbon production increased a thousand-fold if the helium-beryllium collision had a "resonance"—a spike of enhanced fusion probability at a certain energy. Experimental nuclear physicists looked—and the resonance was right where Hoyle predicted [32, 35]. This was viewed as a triumph for the steady-state idea.[9]

Buoyed by this success, and reassured that stars could indeed produce enough carbon-12, Hoyle set out to prove that stars could go on to produce everything else.

He undertook this monumental task along with the Cambridge University astronomers Geoffrey and Margaret Burbidge, and Caltech nuclear physicist William Fowler. In 1957 they summarized their work in a 108-page review paper that is widely known by its authors' initials, B^2FH, and concluded that yes—stellar nucleosynthesis could plausibly account for all the heavy elements [36]. For steady-state believers like Hoyle, this conclusion was another triumph.

Yet ironically, the paper would ultimately come to be seen as support for the Big-Bang model. That's because, as impressive as the B^2FH calculations were, the numbers never quite worked for the lightest elements. In particular, helium does indeed get made by hydrogen fusion in stars, but that

---

[9] It was also one of the first examples of a scientist using what's now called 'anthropic reasoning' to make a testable prediction. That is, Hoyle's arguments hinged on the assumption that some such process must be at play, or we wouldn't be here to ask about it: The universe simply couldn't produce enough carbon to ever give rise to observers.

process couldn't begin to account for helium abundances that astronomers were seeing. This became a real problem for the steady-state theory after 1961, when an influential survey found that the 3-to-1 hydrogen-to-helium ratio was pretty much the same in old stars, young stars, glowing nebulae, interstellar space, distant galaxies—everywhere [37]. This was exactly what you'd expect to see if the hydrogen–helium ratio was primordial—that is, forged in the early universe following a Big Bang—but not at all what you'd expect if the helium had been produced in individual stars with lots of local variation.

Ultimately, these developments would lead to the common-sense compromise that is still the consensus view today: both sides were right, at least when it came to element creation. The light nuclei—hydrogen, deuterium, helium, and a tiny bit of lithium-7—were virtually all made in the Big Bang, as advocated by Gamow, Alpher, and Herman. But every other element was made much later in stars, by the processes described in B^2FH (Fig. 3.5).

In the meantime, an even more serious challenge to the steady-state idea had been brewing since 1955, when the pioneering radio astronomer Martin Ryle of Cambridge University and his student Peter Scheuer published the first reliable survey of "radio stars." These were objects that emitted copious energy at radio wavelengths, but that looked like ordinary stars through a telescope—if they could be seen at all [38]. The nature of these sources was

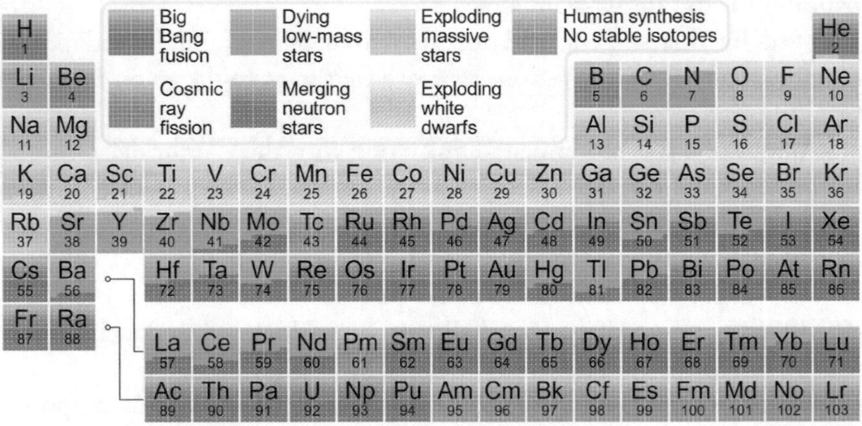

Fig. 3.5 The modern consensus is that the two lightest chemical elements, hydrogen and helium, were almost entirely created during the first few minutes of the Big Bang—as was a tiny bit of element 3, lithium. But all the heavier elements were made much later, via a variety of nuclear processes taking place in stars. (Credit: cmglee/Wikimedia Commons, CC BY-SA 3.0, based on data from Jennifer Johnson, CC BY-SA 4.0)

a mystery at the time. (Some of them would later prove to be the ultrapowerful energy-emitters now known as quasars.) But even so, Ryle and Scheuer were struck by two features of the survey. One was that the radio stars lay outside our galaxy—they were randomly distributed around the sky, with no sign of clustering around the plane of the Milky Way—and were almost certainly at cosmological distances measured in billions of light years. The other was that in every direction Ryle and Scheuer looked, the dimmer, faraway sources substantially outnumbered the brighter, comparatively nearby sources—which was another way of saying that radio stars used to be a lot more common billions of years ago, when those far-away sources had emitted the radio waves that were just now reaching us.

This left Hoyle and his fellow steady-state believers at a loss, since their model insisted that the average distribution of radio stars, galaxies, and every other distant object had to be *constant* over time. "Attempts to explain the observations according to steady-state theories offer little hope of success," Ryle and Scheuer wrote pointedly, "but there seems every reason to suppose that they might be explained in terms of an evolutionary theory." In other words, their findings backed the Big Bang. Subsequent surveys only made the data stronger [25]. So by the early 1960s, most astronomers felt that the steady-state model was on life-support.

All of which set the stage for the third and final rediscovery of Big Bang cosmology, and the chance observation that effectively killed the steady-state idea forever.

3.3   The Big Bang's Afterglow

Steady-state, Big Bang—Robert Dicke didn't much care which model of the universe was right, he was just sick of hearing people argue about it. What the Princeton University physicist *did* care about was testing theories as rigorously as possible. And that most definitely included general relativity [39]. So in the late 1950s, he had embarked on an ambitious program of new, ultra-precise tests of Einstein's theory that went way beyond the bending of starlight around the Sun, and all the other classical tests described in Chap. 2 [40, 41]. To provide a point of comparison, moreover, Dicke had co-authored an alternative now known as the Brans-Dicke model: a non-Einstein theory of gravity that would allow physical laws in the early universe to be quite different than they are today [42].

Dicke and his Princeton associates would carry out many of these precision tests over the coming decades—tests that general relativity would consistently pass. But in the summer of 1964, meanwhile, Dicke had an idea for a new test: find relic radiation from the Big Bang. He apparently didn't know about Alpher and Herman's prediction from 1948. What he was talking about in 1964 was an older idea that the universe might be cyclic, endlessly looping through the same sequence: Big Bang; expand to a maximum size; contract back down to a Big Crunch; bounce outward again into a new Big Bang; repeat. This notion wasn't taken very seriously at the time. But who knew? One Big Bang or many, Dicke thought, all that commotion would surely leave an afterglow, which might just be observable in the microwave region of the radio spectrum.

If so, Dicke was the man to do it. During World War II, when he was part of the MIT team that had developed radar, he had invented a microwave detector known as the "Dicke radiometer," which was (and is) still in widespread use [43]. So he had two young physicists in his lab, Peter Roll and David Wilkinson, build an advanced version of the detector. Roll and Wilkinson were also tasked with building a horn antenna to capture and concentrate the incoming radiation—the name comes from the antenna's resemblance to a boxy cow's horn—as well as a refrigeration system that would maximize the detector's sensitivity by bathing it in liquid helium at 4° Kelvin (4 °C above absolute zero).

It was a fascinating project, Wilkinson later recalled, if only because he and Roll knew exactly what was at stake: If Dicke's idea worked out, he said, "that was very strong evidence for a big bang. There was no way that this heat radiation could be naturally produced in the steady state." [39].

Finally, Dicke assigned a third student, James Peebles, to do the theoretical calculations that would tell the others what to look for. Peebles was dubious at first. As a graduate student, he later recalled, he had been "shocked by the steady state cosmology: they just made this up. But I felt much the same about the relativistic big bang cosmology." Either way, he felt, it was all just airy speculation.

Peebles changed his mind quickly, however: under Dicke, he was learning that cosmology could be a real physical science open to rigorous observation and experiment [39].

Plunging in, Peebles calculated that, after billions of years, cosmic expansion would have cooled the Big Bang emissions down to a few degrees Kelvin. And he realized that nuclear reactions in the Big-Bang fireball would have left us with lots of hydrogen and helium, in ratios that he was calculating quite happily—until he submitted his results to the journal *Physical Review*.

Not new, the editors told him in their rejection letter. These calculations had already been done. In 1948.

And that was how he learned of Gamow, Alpher, and Herman, Peebles later wrote [39]. It was embarrassing—although Dicke, as his mentor, took full responsibility. But Peebles' results were valid even so, and Dicke's team kept going. By early 1965 the theoretical calculations were done. Work on the detector was proceeding nicely. And they were all confident that their work was so far advanced that no one could beat them to the measurement: If the radiation was there, they would see it first.

It was lunchtime on a Tuesday, they all remembered, probably in February 1965. The four of them had gathered in Dicke's office for their weekly team meeting, when Dicke got a telephone call. The three younger physicists thought nothing of it; Dicke got a lot of calls. So they just continued talking among themselves—until they overheard Dicke using words like "horn antenna" and "cold load" (the liquid helium).

"Well boys," Dicke told the now-silent trio when he finally hung up, "we've been scooped."

They had been—and totally by accident. Just 40 km due east of Princeton, at the Bell Labs campus in Holmdel, New Jersey, physicist Arno Penzias and radio astronomer Robert Wilson had spent the previous year trying to make an astronomical instrument out of an old horn antenna originally built for satellite communications (Fig. 3.6). To achieve the extreme sensitivity they were after, Penzias and Wilson had doggedly gone through their apparatus and eliminated every source of interference they could find—radar signals, radio broadcasts, pigeon droppings, everything. But try as they might, there remained a faint microwave hiss that just would not go away. It wasn't due to any glitch in the system they could find. It wasn't coming from the ground; they checked. Nor from the atmosphere; they checked that too. Nor from the Sun. Nor from the plane of the Milky Way galaxy. It was the same day or night. It was the same anywhere in the sky they looked.

No, the hiss was an utter mystery—until a friend, astronomer Bernard Burke of MIT, sent Penzias a draft of one of Peebles' papers on cosmic radiation. And when he and Wilson read it, the meaning of the mysterious hiss suddenly snapped into focus: its intensity corresponded to microwave radiation at a temperature of 3.5 K, give or take a degree, which was well within the range of uncertainty in Peebles' estimate.

The signal just might be cosmic—the afterglow of the Big Bang. Penzias immediately placed a lunchtime phone call to Dicke.

The Princeton group was disappointed at being scooped, to say the least; Dicke later remembered feeling bad for his students, while Peebles, Roll, and Wilkinson remembered feeling bad for *him*. But the news was also exciting: There really was something out there to find! And they could still be the first

Fig. 3.6 Robert Wilson **(left)** and Arno Penzias **(right)** pose with their horn antenna in August 1965, shortly after publishing their pioneering observation of the cosmic background radiation (credit: AIP Emilio Segrè Visual Archives, Physics Today Collection)

to confirm the result—which was critical, since nobody was going to believe the Bell Labs duo until somebody else detected the hiss, too. So Roll and Wilkinson redoubled their efforts to get their detector ready.

Meanwhile, to avoid squabbles over priority, the two groups agreed to publish back-to-back papers in *The Astrophysical Journal*: One from the Princeton team described the theory behind the hiss—that it was heat from the Big Bang—while the other, from Penzias and Wilson, described the observational data [44, 45].

The papers appeared in the July 1 issue of the journal, and were met with-…incomprehension, mostly. And scoffing. Wilkinson later recalled that the steady-staters hated it and the big-bangers didn't believe it [39].

Still, real observational data has a persuasive power that even the most beautiful theory does not. And that data kept accumulating. Later in 1965, Roll and Wilkinson carried their now-finished antenna to the roof of Princeton's Guyot Hall, turned it to the sky, and confirmed Penzias and Wilson's observation—at a different wavelength, to boot: Their apparatus was tuned to a microwave wavelength of 3.2 cm versus Bell Labs' 7.4 cm. This was crucial, because the intensity of radiation at a given temperature varies with wavelength according to a well-known formula known for historical reasons as a blackbody spectrum—and the intensity that the Princeton group found fell right on the curve you'd extrapolate from the Bell Labs detection [46].

So did a very different, but equally compelling measurement from 1966. Another group of astronomers had been searching interstellar space for free-floating molecules of cyanide: a famously poisonous carbon–nitrogen compound that was known to behave like a tiny antenna tuned to 0.26-cm microwaves. What they found was that these interstellar molecules were absorbing and re-emitting photons at a temperature consistent with 3 K—exactly what you'd expect if they were being bathed in cosmic radiation at that temperature [47, 48].

By the 1970s, the steady accumulation of observations like these had led to an almost universally acceptance that the cosmic microwave background (CMB) was real—and that the only reasonable explanation for it was the Big Bang theory.

The CMB's discovery would earn Penzias and Wilson a share of the 1978 Nobel Prize in physics—an honor that the Princeton group hailed as richly deserved, although Peebles still wishes that the Nobel committee had also named Dicke, who died in 1997. Wilkinson would devote the rest of his career to studying the CMB in every way he could—from the ground, from high-altitude balloons, and eventually from space. If not for his untimely death from cancer in 2002, he might have shared in the 2006 physics Nobel, which recognized the findings of the Cosmic Background Explorer satellite that he had helped design. Peebles would also make cosmology his life's work, coming up a host of theoretical insights that *would* earn him a share of the Nobel physics prize in 2019.

Fred Hoyle, for his part, would reject the Big Bang idea until the day he died in 2001. But after the CMB discovery, even he could see that the steady-state alternative was no longer viable in anything like its original form. Hoyle began to investigate a series of Big-Bang alternatives that most other astrophysicists found increasingly bizarre—an obsession that may have cost him a share in the 1983 Nobel that went to Fowler for providing the data that underlay the B^2FH paper on stellar nucleosynthesis.

Gamow, by contrast, felt vindicated by the CMB discovery—if perhaps a little miffed that others were getting all the attention. As he was heard to declare at one meeting in 1967, he had lost a penny. Penzias and Wilson had found a penny. Was it *his* penny? [39]

Abbé Lemaître, by contrast, was both gratified and gracious. News of Penzias and Wilson's discovery reached him in Belgium in June 1966, as the 71-year-old president of the Pontifical Academy of Sciences lay dying of leukemia. In three days he would be gone, writes science historian Simon Mitton [49]. Yet gravely ill though he was, "Lemaître lucidly praised this news, which confirmed the explosive genesis of our universe."

References

1. Kragh H (2008) The origin and earliest reception of big-bang cosmology. Publ Obs Astron Beogr 85:7–16
2. Eddington AS (1931) The end of the world: from the standpoint of mathematical physics*. Nature 127:447–453. https://doi.org/10.1038/127447a0
3. Lemaître G (1931) The beginning of the world from the point of view of quantum theory. Nature 127:706. https://doi.org/10.1038/127706b0
4. Lemaître AG (1931) Contributions to a British association discussion on the evolution of the universe. Nature 128:704–706. https://doi.org/10.1038/128 704a0
5. Eddington AS (1920) The internal constitution of the stars. Observatory 43:341–358
6. Gingerich O (1988) Shapley's impact. In: Proceedings of IAU Symposium No. 126. Kluwer Academic, Cambridge, MA, p 23
7. Wayman PA (2002) Cecilia payne-gaposchkin: astronomer extraordinaire. Astron Geophys 43:1.27–1.29. https://doi.org/10.1046/j.1468-4004.2002.431 27.x
8. Gingerich O (1982) Obituary–payne-gaposchkin cecilia. Q J R Astron Soc 23:450
9. Gingerich O (2001) The most brilliant. PhD thesis, Ever Written in Astronomy. p 3
10. Payne CH (1925) Stellar atmospheres; a contribution to the observational study of high temperature in the reversing layers of stars
11. Payne CH (1925) Astrophysical data bearing on the relative abundance of the elements. Proc Natl Acad Sci 11:192–198. https://doi.org/10.1073/pnas.11.3.192
12. Chernin AD (1995) George gamow and the big bang. Space Sci Rev 74:447–454. https://doi.org/10.1007/BF00751431
13. Gamow G (1928) The quantum theory of nuclear disintegration. Nature 122:805–806. https://doi.org/10.1038/122805b0
14. Bethe HA (1939) Energy production in stars. Phys Rev 55:434–456. https://doi.org/10.1103/PhysRev.55.434
15. Friedmann A (1924) Über die Möglichkeit einer Welt mit konstanter negativer Krümmung des Raumes. Z Für Phys 21:326–332

16. Friedmann A (1999) On the possibility of a world with constant negative curvature of space. Gen Relat Grav 31:2001–2008
17. Gamow G (1946) Expanding universe and the origin of elements. Phys Rev 70:572–573. https://doi.org/10.1103/PhysRev.70.572.2
18. Peebles PJE (2014) Discovery of the hot big bang: what happened in 1948. Eur Phys J H 39:205–223. https://doi.org/10.1140/epjh/e2014-50002-y
19. Gamow G (1948) The evolution of the universe. Nature 162:680–682. https://doi.org/10.1038/162680a0
20. Alpher RA, Bethe H, Gamow G (1948) The origin of chemical elements. Phys Rev 73:803–804. https://doi.org/10.1103/PhysRev.73.803
21. Kragh H (2001) Nuclear archaeology and the early phase of physical cosmology. pp 252–157
22. Alpher RA, Herman R (1948) Evolution of the universe. Nature 162:774–775. https://doi.org/10.1038/162774b0
23. Kragh H (1996) Cosmology and controversy. Princeton University Press, Princeton, NJ
24. Bondi H (1990) The cosmological scene 1945–1952. In: Modern cosmology in retrospect. Cambridge University Press, p 189
25. Kragh H (2012) Quasi-steady-state and related cosmological models: a historical review. Astro-Ph Phys ArXiv12013449
26. Bondi H, Gold T (1948) The steady-state theory of the expanding universe. Mon Not R Astron Soc 108:252. https://doi.org/10.1093/mnras/108.3.252
27. Hoyle F (1948) A new model for the expanding universe. Mon Not R Astron Soc 108:372. https://doi.org/10.1093/mnras/108.5.372
28. Hoyle F An online exhibition. Retrieved March 2, 2020, from https://www.joh.cam.ac.uk/library/special_collections/hoyle/exhibition/radio/
29. Hoyle F (1951) The nature of the universe. First American. Harper, New York
30. Kragh H (2013) What's in a Name: history and meanings of the term "Big Bang." ArXiv E-Prints 1301: arXiv:1301.0219
31. Hoyle F (1946) The synthesis of the elements from hydrogen. Mon Not R Astron Soc 106:343. https://doi.org/10.1093/mnras/106.5.343
32. Hoyle F (1954) On nuclear reactions occuring in very hot stars. I. the synthesis of elements from carbon to nickel. Astrophys J Suppl Ser 1:121. https://doi.org/10.1086/190005
33. Hoyle F, Fowler WA, Burbidge GR, Burbidge EM (1956) Origin of the elements in stars. Science 124:611–614. https://doi.org/10.1126/science.124.3223.611
34. Salpeter EE (1952) Nuclear reactions in stars without hydrogen. Astrophys J 115:326–328. https://doi.org/10.1086/145546
35. Dunbar DN, Pixley RE, Wenzel WA, Whaling W (1953) The 7.68-Mev state in C12. Phys Rev 92:649–650. https://doi.org/10.1103/PhysRev.92.649
36. Burbidge EM, Burbidge GR, Fowler WA, Hoyle F (1957) Synthesis of the elements in stars. Rev Mod Phys 29:547–650. https://doi.org/10.1103/RevModPhys.29.547

37. Osterbrock DE, Rogerson JB Jr (1961) The helium and heavy-element content of gaseous-nebulae and the sun. Publ Astron Soc Pac 73:129. https://doi.org/10.1086/127637

38. Ryle M, Scheuer PAG (1955) The spatial distribution and the nature of radio stars. Proc R Soc Lond Ser A 230:448–462. https://doi.org/10.1098/rspa.1955.0146

39. Peebles PJE, Page LA Jr, Partridge RB (2009) Finding the Big Bang. Cambridge University Press

40. Dicke RH (1959) New research on old gravitation. Science 129:621–624 https://doi.org/10.1126/science.129.3349.621

41. Peebles PJE (2017) Robert Dicke and the naissance of experimental gravity physics, 1957–1967. Eur Phys J H 42 https://doi.org/10.1140/epjh/e2016-70034-0

42. Brans C, Dicke RH (1961) Mach's principle and a relativistic theory of gravitation. Phys Rev 124:925–935. https://doi.org/10.1103/PhysRev.124.925

43. Dicke RH (1946) The measurement of thermal radiation at microwave frequencies. Rev Sci Instrum 17:268–275. https://doi.org/10.1063/1.1770483

44. Dicke RH, Peebles PJE, Roll PG, Wilkinson DT (1965) Cosmic black-body radiation. Astrophys J 142:414–419. https://doi.org/10.1086/148306

45. Penzias AA, Wilson RW (1965) A measurement of excess antenna temperature at 4080 Mc/s. Astrophys J 142:419–421. https://doi.org/10.1086/148307

46. Roll PG, Wilkinson DT (1966) Cosmic background radiation at 3.2 cm-support for cosmic black-body radiation. Phys Rev Lett 16:405–407. https://doi.org/10.1103/PhysRevLett.16.405

47. Field GB, Herbig GH, Hitchcock J (1966) Radiation temperature of space at $\lambda 2.6$ mm. Astron J 71:161–161. https://doi.org/10.1086/110071

48. Thaddeus P, Clauser JF (1966) Cosmic microwave radiation at 2.63 mm from observations of interstellar CN. Phys Rev Lett 16:819–822. https://doi.org/10.1103/PhysRevLett.16.819

49. Mitton S (2016) Georges lemaître: life, science and legacy. ArXiv E-Prints 1612: arXiv:1612.03003

4

Behind the Veil

It's a fundamental fact found in every astronomy textbook: The further *out* we look into space, the further *back* we look into time.

When we gaze at the moon, for example, we're seeing it as it was about 1½ s ago—the time it takes for sunlight to bounce off its surface and cross the intervening 384,400 km to our eyes. We likewise see the Sun as it was about 500 s ago, or 8 min and 20 s: the time it takes for photons moving at the speed of light to cross the 150-million-km radius of Earth's orbit. It's the same story with the planets, whose light-transit times are measured in hours; and the nearby stars, which we're seeing as they were a few years or centuries ago.

These delays are rarely long enough for things to change significantly—not even when we look at the Andromeda galaxy, which we're seeing by the light it emitted some 2.5 million years ago, shortly before a species now known as *homo habilis* started wandering the plains of Africa. Astronomical objects tend to be *big*, and to evolve very slowly on any human timescale.

But look out far enough and the changes *do* start to matter. The furthest galaxies detected by the Hubble Space Telescope are noticeably different from their descendants today because they are infants: astronomers are seeing them at a time when the first galaxies were just beginning to form, less than a billion years after the Big Bang. And the faint hiss of the Cosmic Microwave Background (CMB) gives radio astronomers a view of the universe from long before that, back when there were no galaxies and the 380,000-year-old universe contained nothing but rapidly cooling hydrogen and helium gas.

© The Author(s), under exclusive license to Springer Nature
Switzerland AG 2022
M. M. Waldrop, *Cosmic Origins*,
https://doi.org/10.1007/978-3-030-98214-0_4

For better or worse, however, this is as far as astronomy's time machine can take us. Before the creation of the CMB, the universe was a haze of ionized matter that scattered and randomized any photon that tried to pass through it; looking further back is like trying to look beneath the surface of the Sun.

So cosmologists are left with two options. Just as with the sun, they can learn an enormous amount by studying the CMB itself and all the space in between; those investigations will be the subject of Chap. 5. Or, as we'll discuss in this chapter, cosmologists can try to infer what went on behind the veil through a combination of theory and extrapolation from lab experiments.

What we'll find is that science still hasn't found an answer the most fundamental question: where did the universe come from? But it *has* shown us some radically new possibilities.

4.1 Grand Unified Cosmology

4.1.1 The Two Standard Models

Quite aside from transforming our understanding of cosmic origins, the discovery of the cosmic microwave background transformed the field of cosmology. After 1965, it was a niche subject no longer.

Partly this was because the discovery upended cosmology's reputation as a realm of hopelessly vague speculation, and proved that it had become a normal science with predictions that could be precisely formulated and tested [1]. That newfound respectability, in turn, sparked a huge wave of interest in the subject from astronomers and physicists alike.

But the field's surging popularity was also driven by a historical coincidence. During the 1950s through the 1970s, the same period that saw astronomers coalesce around the Big Bang as their "standard model" of cosmic origins, physicists were developing a standard model of their own—a theory that started with James Clerk Maxwell's nineteenth century unification of electricity and magnetism, which we described in Chap. 2, and expanded it to include both the strong forces that dominate the atomic nucleus, and the weak interactions responsible for certain forms of radioactivity. In fact, this standard model included all the forces known to exist in nature except for gravity, which was already covered by Einstein's general relativity [2] (Fig. 4.1).

The trick, as always, was to test this theory, which would require smashing particles together at energies far beyond any existing accelerator. The drive to fill that gap would eventually lead to the construction of new-generation

Standard Model of Elementary Particles

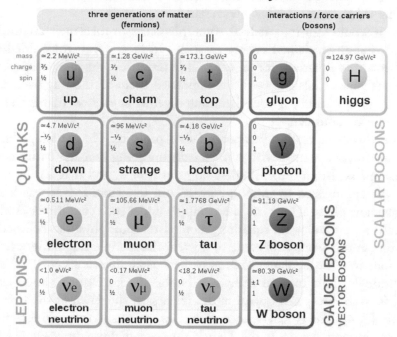

Fig. 4.1 In the standard model of particle physics, every non-gravitational force is carried from one particle to the next by one of the *gauge (or vector) bosons* in the fourth column. The gluons carry the strong forces of quantum chromodynamics, the photon carries the electromagnetic force, and the W and Z particles carry the weak force. Each of these force-carrier particles contains one quantum unit of internal angular momentum, or spin. The Higgs boson in the fifth column is the "frozen" field that gives the other particles mass. It is an example of a *scalar* boson that has a spin of zero. On the left are the building blocks of matter. Known generically as *fermions*, these particles vary in their mass and electric charge, but always carry spin ½. Each of these fermions also has an antiparticle (not shown) with the same mass and spin, but opposite electric charge. The rows in this chart indicate which forces the fermions respond to (in addition to gravity, which affects them all.) The *quarks* in the first two rows respond to all three forces: strong, weak, and electromagnetic. The much less massive *leptons* at the bottom—the name comes from *leptos*, the Greek word for "slim"—ignore the strong force. But the electron and its partners in the third row respond to both the weak and electromagnetic forces, while the neutrinos in the last row respond only to the weak force. The columns in the fermion part of this chart indicate the three known *generations* of fermions. It's not clear why nature choses to have all three, since the fermions in the first column comprise all the ordinary matter we see here on Earth or in the heavens, and the other two generations seem to be identical to the first in every way except mass. But these heavier relatives turn up regularly in cosmic-ray and accelerator experiments. So a major goal of any grand unified theory is to explain why all these particles and forces exist—but no others. (MissMJ, Cush/Wikimedia Commons, CC BY 3.0)

accelerators like the current high-energy champion: the Large Hadron Collider (LHC) near Geneva, Switzerland.

But it would also lead at least some physicists to turn their attention to the Big Bang, which had energy to spare. In fact, if you took Friedmann and Lemaître's equations for an expanding universe—the ones derived from Einstein's general relativity—and extrapolated them all the way back to the beginning, you'd find particles slamming together at energies higher than any number you could name.

True, this extrapolation was potentially a problem: The equations insisted that the universe started with a *singularity*, meaning that the entire cosmos was compressed to a dimensionless point at $t = 0$, and that all the matter in it was compressed to infinite density and temperature.[1] It was hard to imagine how physical laws could even apply under such conditions. And that included the cosmological equations themselves: Most physicists believed that quantum effects would intervene to prevent that cosmic singularity once the equations were extrapolated back past the *Planck time* at $t = 10^{-43}$ s—an astonishingly short interval named in honor of Max Planck,[2] the German physicist who discovered quantum discreteness at the dawn of the twentieth century [3, 4]. The closely-related *Planck energy* was likewise an estimate of the cosmic temperature at the Planck time—and thus, the average energy of particle collisions. This energy worked out to an equally astonishing 10^{19} billion electron Volts (GeV).[3]

Precisely what happened before the Planck time was (and is) a gigantic question mark. Did Einstein's gravity and the newly discovered unified theories of particle physics somehow merge into the ultimate unification—a.k.a. the Theory of Everything? Did time and space themselves somehow dissolve in this Planck era, falling apart into primordial fragments that were even more basic than space–time?

---

[1] The equations also insisted that there was nothing before t = 0, simply because you couldn't extend the coordinates through the singularity to reach earlier times. It was a bit like asking what lies south of the South Pole.

[2] In 1899, when he was first developing quantum theory, Planck proposed these units as a way to measure length, time, energy, and every other physical quantity in terms of fundamental constants of nature like the speed of light, and Newton's constant of gravitation. Planck argued that such a system was "natural," having no reference to human constructs like the meter, and thus "necessarily retaining their meaning for all times and for all civilizations, including extraterrestrial and non-human ones." Crucially, though, he could make this work only by including a new fundamental constant that he had proposed earlier in the same paper. Today, of course, it is known as Planck's constant, and is central to the quantum revolution that Planck set in motion.

[3] Researchers doing cosmological calculations often (but not always) measure temperatures in units of energy. The concepts are equivalent: the temperature of a sample of matter is proportional to the average kinetic energy of the particles that comprise it. But the numbers tend to be more convenient. For example, an energy of one billion electron Volts—1 GeV—corresponds to a temperature of roughly 11 trillion degrees Kelvin.

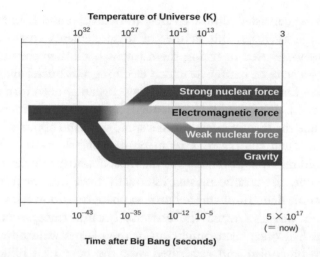

Fig. 4.2 The unified theories of particle physics developed over the past half-century revealed that the contents of the universe underwent profound changes as the cosmos expanded and cooled. Many questions remain about the very earliest instants, sketched here on the left. But it's thought all the particles were essentially alike then, as were the four fundamental forces between them. And there is broad consensus that after $t = 10^{-43}$ s, when gravity diverged and became much weaker, the other three forces crystalized and took on their present form in a series of dramatic phase transitions [113]

Fortunately, researchers could make plenty of progress without waiting for all the answers to such questions. The Friedmann-Lemaître cosmological equations could presumably be trusted back *to* the Planck time. And that left plenty of scope for physicists to explore what their standard model could tell them about the early universe—and in the process, to go far further back towards $t = 0$ than anyone could have imagined just a few years previously (Fig. 4.2).

4.1.2 Cosmic Phase Transitions

It turned out that a particularly useful way to tackle this problem was to think about the early universe as a series of *phase transitions*—the cosmic analog of what happens when solid ice melts into liquid, or when water boils away into steam [5].

The Recombination Epoch: 380,000 Years

A good example is the formation of the cosmic microwave background. As we discussed in Chap. 3, the matter that emerged from the first few minutes of the Big Bang consisted almost entirely of hydrogen and helium in a ratio of three to one by mass. But neither of these elements emerged as a gas, which is how we're used to seeing them today. Back then they existed only as a *plasma*—a state of matter in which electrons and nuclei are so hot that they can't even form atoms. They would have instantly been torn apart again if they even tried. As the universe expanded, however, this primordial plasma cooled, just like the mist coming out of a spray can. And about 380,000 years after $t = 0$, when the plasma's temperature fell below 10,000° K or so, electrons could finally start binding to the most abundant nucleus, hydrogen.

At this point, the cosmic plasma essentially froze into neutral hydrogen atoms, plus some deuterium and helium—a phase transition known as *recombination*. This was also when the universe became transparent, since the photons that had once been caught up in the plasma were now able to fly free—and to be cooled and redshifted over the next 13.8 billion years of cosmic expansion into the CMB we see today.

The Nucleosynthesis Epoch: 1 s

Another cosmic phase transition discussed in Chap. 3 was the era of nucleosynthesis, which was first analyzed by Gamow, Alpher, and Herman in 1948 and then rediscovered by Peebles in the 1960s. This transition had happened 3800 centuries earlier than the first, in the 15 min that started around $t = 1$ s. And it unfolded at cosmic temperatures a million times higher [6]. But the process was much the same: Just as atoms couldn't exist before the CMB transition, heavy nuclei couldn't exist before this one. Prior to $t = 1$ s, protons and neutrons would have been outnumbered about a billion to one by light particles—mostly photons, but also hordes of neutrinos, electrons, and positrons (anti-electrons). The resulting firestorm would have instantly shredded any deuterium or helium that tried to form.

After about $t = 1$ s, though, as temperatures fell below about 10 billion K, photons wouldn't have enough energy to disrupt things anymore. And after another minute or so, neither would the electrons and positrons: they would mostly just annihilate each other and vanish into more photons. As a result, the protons and neutrons would be free to spend the next quarter-hour in a frenzy of thermonuclear fusion, cooking up today's mix of hydrogen, deuterium, and helium. And after that, when temperatures fell past the point where those reactions could be sustained, the mix would essentially freeze into place—preserving the isotope ratios we see today.

4.1.3 The Quark Epoch: 10^{-5} s

The nucleon soup that existed up until the nucleosynthesis phase transition was about as close to $t = 0$ as anyone could get with the nuclear physics that was known in 1948, or even in the early 1960s. But that barrier would vanish less than a decade later with the development of *quantum chromodynamics* (QCD): the first successful theory that explained what protons and neutrons were, and what was going on with the strong forces [7]. This theory was also, not incidentally, a critical piece of what would soon become known as the standard model.

QCD revolved around *quarks*: whimsically named entities that were supposed to be the building blocks for protons and neutrons, as well as for a host of related, but short-lived particles that had started turning up in physicists' particle collision detectors during the 1950s. The members of this particle zoo were collectively known as *hadrons*, from the Greek word for "stout," and at first seemed to be united mainly by the fact that they all responded to the strong force; in most other respects, they were an incomprehensible mess. And the more powerful the accelerators became, the more of them there seemed to be.

But the quark idea revealed the hidden order in this chaos. The hadrons made perfect sense if you simply assumed that they were all built from quarks in various combinations; that the quarks came in three different *colors* (a whimsical term that referred to an analog of electric charge, not color vision); and that these tri-colored building blocks were bound together by eight photon-like force particles dubbed *gluons*. The resulting equations were like a generalization of *quantum electrodynamics,* the theory that describes the behavior of photons and the electromagnetic force at a subatomic level. So, given the multicolored quarks, the name quantum chromodynamics was almost inevitable. And that playful phrase was catchy enough that physicists were soon using it in serious scientific papers—at least some of which were attempts to use QCD for cosmology, and to figure out where that nucleon soup came from before $t = 1$ s.

In human terms, of course, 1 s is just a single heartbeat, a barely noticeable fraction of a lifetime. But in the early universe that first cosmic heartbeat was a vast expanse of time, full of incident, change, and drama at exponentially rising energy scales.

Reaching the quark epoch turned out to be a prime example: To get there, you had to imagine the cosmic clock ticking down from $t = 1$ s to roughly $t = 10^{-5}$ s, while the cosmic temperature soared past 10^{12} (one trillion) degrees. Under those conditions, the QCD calculations showed, not even protons or neutrons could exist anymore. Push back any further and the nucleon soup would undergo a phase transition much like the first two we mentioned; in effect, the protons and neutrons would melt and release their building blocks to form a free-flowing *quark-gluon plasma*. Or, if we were to turn around for a moment and look forward from this point, we would find protons and neutrons crystallizing out of the rapidly expanding plasma like tiny hailstones. (In recent years, accelerators such as the Large Hadron Collider in Geneva, Switzerland, have been able to recreate this transition by colliding heavy nuclei such as gold, and thereby verify predictions about the resulting quark-gluon plasma as it expands and crystallizes into particles inside the detectors [8].)

4.1.4   The Electroweak Epoch: 10^{-12} s

Pushing further back toward $t = 0$, the next phase transition involved the other big piece of the standard model: an *electroweak* theory that unified electromagnetism with the weak interactions [9–12].

The electroweak theory developed in the 1950 and 1960s more or less in parallel with QCD—and indeed, used many of the same mathematical ideas. But it arguably required a much more radical leap of imagination. For starters, the unification equations would work only if the photon of electromagnetism were joined not by *one* new force particle for the weak interactions, but three—all of which would have to be 80 or 90 times the mass of the proton.[4] The existence of these force particles wouldn't be experimentally confirmed until 1983 [13].

Worse, though, the theory also insisted that electrons had an intimate mathematical relation to *neutrinos*: particles that were already known to respond to the weak interactions in much the same way as electrons. The problem was that, weak interactions aside, these particles were utterly different. Electrons have mass—some 0.511 million electron volts (Mev) in the energy units[5] favored by physicists, or roughly 1/2000 the mass of the

---

[4] The three were named the W^+, W^-, and Z^0. The first two have positive and negative electric charges, respectively, and were named after the weak interaction. The third has zero charge, and was named after…well, zero. As for their mass: having your force carried by a very massive particle is equivalent, in the quantum realm, to saying that the force is very short range. And the range of the weak interaction range was so short that for a long time, physicists thought it was zero—meaning that the weak force appeared to act only when particles were in direct contact.

[5] Again, this is a matter of convenience. Atomic-scale particles are ridiculously small by any everyday measure. So if you tried to gauge them in everyday terms using, say, the metric system, you'd get an electron mass that was awkward, to say the least: 9.1×10^{-31} kg. The physicists' units yield much smaller numbers by relying on Einstein's mass-energy equivalence relation, $E = mc^2$. The

proton. Neutrinos have zero mass and always move at the speed of light.[6] Electrons also carry an electric charge, and thus respond to the electromagnetic force. Neutrinos do neither; since they respond only to the weak force, they almost never interact with everyday matter—and indeed, can sail right through the Earth as though it wasn't there. They are about as ghostly as particles can get.[7]

The key insight that made the intimate relation between electrons and neutrinos comprehensible was that their near-identical response to the weak force was actually the relic of a much deeper *symmetry* between them. And by adding a certain quantum field to the equations, physicists discovered, this symmetry could be made manifest—meaning that the two particles would enter into the equations in exactly the same way.

But what did this field actually *do* in the real world? Nothing, usually. Under any conditions found in our present-day universe—including at the core of even the hottest exploding star—the new field was "frozen" in a certain mathematical sense.[8] The result wasn't a physical crystal drifting through the universe like a snowflake. Instead, the field was locked into a single, fixed value that was the same at every point in space. And that, in turn, produced a subtle reorganization of the equations that "broke" the symmetry, gave mass to the electrons and three of the four force particles (while leaving the photons and neutrinos massless), and ended up describing exactly the particles and interactions we observe.

However, this didn't mean that the frozen field was forever beyond the reach of observation. Since every field in the quantum realm corresponds to a particle, there had to be a particle associated with this one, as well—a really massive particle dubbed the Higgs boson. The Higgs boson's 2012 discovery at the LHC—with a 125 GeV mass that was more or less as predicted—would confirm the last remaining piece of the standard model, and would be greeted with worldwide fanfare (and Nobel prizes) [14, 15]. In effect, the

---

electron-Volt (eV) unit also happens to be convenient when you're working with particle accelerators: 1 eV is the amount of energy picked up by a particle with one unit of electric charge—an electron or proton, for example—when accelerated by a potential difference of 1 V.

[6] Or at least, that's what physicists thought in the 1960s. It's now known that neutrinos have an infinitesimally small, but non-zero mass—which means that they always travel at very *close* to the speed of light, but never quite get there.

[7] Neutrinos do have the same internal angular momentum, or *spin*, as an electron. In quantum units, they both have spin ½.

[8] In technical papers, a field like this is known as a *scalar* field, meaning first, that it can be defined by a single (complex) number at every point in space; and second, that the associated particle has a spin of 0. Also, the technical term isn't "frozen"; rather, the field is said to have a non-zero *vacuum expectation value*. But as we'll see in a moment, the concepts are very close.

LHC was finally able to produce particle collisions at a high enough energy to melt the frozen field ever so slightly, and shake a Higgs boson loose.

And of course, this melting process is exactly what happened on a cosmic scale in the early universe. By 1974, a number of physicists had shown that this "freezing" analogy wasn't *just* an analogy [16–18]. If you ran the cosmic countdown clock past about $t = 10^{-12}$ s, so that the temperatures exceeded some 10^{15} K, the Higgs field would quite literally melt and restore the symmetry it had broken. Above that temperature, electrons would cease to have mass and would start to look virtually identical to their companion particles, the neutrinos. Likewise with photons and the other three electroweak force particles: Their masses would vanish, and their similarities would become manifest.

Baryogenesis?

This electroweak era may also have encompassed an additional phase transition known as *baryogenesis*, a name derived from *barys*, the Greek word for "heavy." It's the event that allowed everything we see around us to be made of matter instead of antimatter.

We take this simple fact for granted, but from the particle physics perspective it's quite mysterious. Basic quantum theory suggests that matter and antimatter ought to be like mirror images of one another (albeit with their charges reversed and their internal clocks running backward). And if that's the case, then the early universe should have brought forth both types in equal numbers: Electrons and positrons, neutrinos and antineutrinos, quarks and antiquarks, each with the same abundance as its antiparticle twin. As the universe cooled, moreover, those twins should have annihilated each other on contact, leaving behind only pairs of photons. By now the universe ought to contain nothing *but* photons.

Yet here we are—which is why there must have been a baryogenesis event that tilted the balance and allowed the ordinary stuff to dominate. It wouldn't need to be much of a tilt: observations suggested that matter particles initially outnumbered their antiparticle counterparts by only about one part in 10 billion—a number so small as to be humbling. But even so, it was a number that made our existence possible, and that needed an explanation.

In 1967, the Soviet physicist Andrei Sakharov provided one. For baryogenesis to happen in the rapidly expanding early universe, he showed, the particle equations would have to contain fields that broke the particle-antiparticle symmetry in certain subtle ways [19]. Sakharov wasn't able to do much more with this idea, since he was preoccupied with his increasingly vocal opposition to the Soviet regime. But a decade later, other physicists revived it in the context of the standard model and found that the electroweak interactions allowed for just the kind of symmetry-breaking Sakharov had postulated [20, 21]. There were (and are) plenty of uncertainties, and it's far from clear even now whether the electroweak fields can fully account for that one part in 10 billion figure. But recent findings in neutrino physics appear to strengthen the case considerably [22].

4.1.5 The Grand Unification Epoch: 10^{-36} s

As the 1970s went on, the phrase "standard model" became an increasing common shorthand for QCD and the electroweak theory taken together. Between them, after all, these two sets of equations seemed to give a complete account of all the known forces besides gravity.[9] And of course, two theories had collectively given physicists a way to figure out the cosmic timeline all the way back to $t = 10^{-12}$ s.

As impressive as that feat was, however, theorists were already hoping to unify their own unifications—and in the process, push the cosmic countdown a *lot* closer to $t = 0$.

Grand Unified Theories

Even in the 1970s, theorists had plenty of reasons to think that the standard model couldn't possibly be the whole story. For one thing, it seemed so...arbitrary. Why did nature choose *this* particular set of particles, and *this* particular set of strong, electromagnetic, and weak forces to guide them? Why *this* particular unification scheme and not some equally plausible alternative?

For another thing, QCD and the electroweak theory had a sizeable overlap, with quarks responding to electromagnetism and the weak force as much as to the gluons. And more than that, the equations that governed the various forces seemed to have deep similarities—a common mathematical structure that made them look like different aspects of a much larger whole.[10] It was as if the standard model were crying out for a higher-order unification.

---

[9] The phrase "standard model" steadily gained in popularity and credibility each time another of the equations' predictions was confirmed at the big particle accelerators—the Higgs discovery being only the last and most spectacular. By now, every particle and every interaction in the standard model has been abundantly verified—and conversely, nothing has been seen (yet) at the big accelerators that isn't part of the standard model. As we'll see in the next chapter, however, there are other aspects of the universe that do *not* fit the standard model, starting with dark matter and dark energy.

[10] This is most obvious when you look at the symmetries in each of the theories. The simplest is Maxwell's electromagnetism, which obeys a symmetry that mathematicians denote $U(1)$. The technical details get deep into the quantum description of charged particles. But it's not too far wrong to say that the 1 in $U(1)$ is ultimately tied to the fact that there is only one type of electric charge (albeit

And indeed, achieving that goal turned out to be comparatively straightforward [23–25]. By 1974, physicists had begun to come up with several good prospects for an all-encompassing *Grand Unified Theory*, or GUT. True, there was no experiment we could do here on Earth to tell these GUTs apart. At everyday energies they each gave exactly the same predictions as the standard model—which is what they'd been deliberately constructed to do, after all. The candidates differed mainly in their requirements for new frozen fields, known generically as *scalars*, and in their predictions for new, ultra-high-energy interactions.

Still, physicists had powerful, if indirect evidence that they were on the right track with grand unification.

Converging on GUT

This clue grew out of efforts to take the weak, electromagnetic, and QCD interactions—all the forces covered by the standard model—and calculate what each one predicted for particles colliding at extreme energies. The answer was by no means obvious from the basic equations; as always in the quantum realm, smashing the particles together at higher and higher energies meant probing the forces at smaller and smaller distances. This quickly brought both relativity and quantum uncertainty into play,[11] and posed a calculational challenge for the theories that could be tackled only with specialized mathematical tools.[12]

---

with values that can be either positive or negative), and only one type of particle that connects those charges with the electromagnetic force—namely, the photon.

For the electroweak unification, physicists expanded that symmetry to *SU(2) × U(1)*, where the 2 encoded (among other things) the electron-neutrino pairing. And when they worked through the mathematical details, the number of new force particles required for the *SU(2)* part turned out to be 2^2-1, or 3.

For quantum chromodynamics physicists likewise employed a symmetry called *SU(3)*, which encoded the fact that each quark carried one of three "color" charges. By the same logic as before, the number of new force particles required by SU(3) turned out to be 3^2-1, or 8; thus the eight gluons. However, the quarks also feel both electromagnetism and the weak force, so they are actually governed by the product of all three symmetries. The result is *SU(3) × SU(2) × U(1)*—which is the symmetry of the standard model.

The structural similarities that seem obvious in the names were just as obvious in the mathematics; thus the motivation for a Grand Unified Theory (GUT) that could tie them all together. This meant finding a larger symmetry that could incorporate *SU(3) × SU(2) × U(1)* in a natural way (and in the process, explain why nature chose these particular symmetries, as opposed to the multitude of equally plausible possibilities). There turned out to be lots of candidates—classic examples such as *SU(5)* and its slightly larger cousin *SO(10)*, but many, many more.

[11] Among other things, quantum uncertainty allows pairs of particles and their antiparticles to spring into existence out of nothing. These entities will annihilate each other and disappear almost instantaneously. But during their brief existence they can have a large effect on short-distance forces.

[12] Easily the most powerful of these tools was the concept of the *renormalization* group, which provided a mathematically rigorous way to relate long-distant forces to short-distance forces while

But by the early 1970s, even as the various GUTs themselves were being invented, these tools had developed to the point where physicists could show that both the electromagnetic and weak forces got stronger at very short distances, in roughly the same way that ordinary electrostatic forces do.[13] Using these same techniques, however, physicists found that the strong forces described by QCD got *weaker* at short distances [26, 27]. It was as if quarks were connected by something like a rubber band: The forces between them were large when the quarks were far apart and the band was stretch tight, but negligible when the quarks were close, and the band was relaxed.[14]

Crucially, however, this meant that, as the particle collision energies went up, the forces would converge: The weak and electromagnetic interactions would get stronger, while the immense nuclear forces described by QCD would get weaker. And when these calculations were extrapolated to very, very high energies, the three curves came together at about 10^{16} GeV.[15] Presumably this was the energy scale at which all the frozen fields in the true GUT would melt, and grand unification would become manifest (Fig. 4.3).

A Plague of Monopoles

As reassuring as this threefold convergence was, however, the numbers also gave people pause. Pushing the cosmic story back to the grand unification epoch was going to require a truly breathtaking leap. In terms of time, it meant extrapolating Einstein's equations by another 26 orders of magnitude, from the electroweak transition at roughly $t = 10^{-12}$ s down to a GUT transition at something like $t = 10^{-38}$ s. Or in terms of temperature, it meant extrapolating from the electroweak energy scale of 100 GeV up to

---

taking full account of both relativity and quantum mechanics. Renormalization group methods are applicable in many fields, including the theory of condensed matter. But it wasn't until the 1950 and 1960s that they began to make themselves felt in particle physics.

[13] Whether it's like charges repelling or unlike charges attracting, the electrostatic force goes as the inverse square of the distance—which means that it technically become infinitely strong at zero distance.

[14] This was also why it's impossible to isolate a free quark. Try to pull one out of a proton or a neutron, and the "rubber band" would eventually stretch to the point of breaking. But the resulting energy release would produce a quark-antiquark pair, which would instantly separate and attach to the broken ends, producing normal-looking hadrons with the quarks bound inside.

[15] To be technically accurate, what converged were the forces governed by each piece of the standard model's $SU(3) \times SU(2) \times U(1)$ symmetry, using the notation from an earlier footnote. This isn't quite the same as saying that the weak and electromagnetic forces strengthened independently, since the $SU(2) \times U(1)$ part mixed the two. But the idea was the same. Also, the intersection at 10^{16} GeV was only approximate—although still surprisingly close. And if physicists included the exotic, but theoretically beautiful principle known as supersymmetry, the intersection was exact.

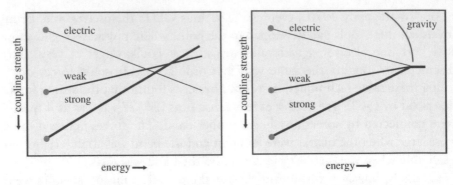

Fig. 4.3 Calculations in the 1970s showed that, as particles collided with higher and higher energies—and thus probed the forces between them at shorter and shorter distances—the strength of the three fundamental forces encompassed by the standard model became more and more alike. (Note that what's plotted here is the inverse of the coupling strength, so that a line sloping upward is actually getting weaker with energy.) **Left:** The three lines came close to meeting at about 10^{16} GeV. **Right:** When the equations incorporated a hypothetical, but mathematically beautiful principle known as *supersymmetry*, the convergence became exact [112]

10^{16} GeV—a factor of 100 trillion. That was still three orders of magnitude below the Planck energy of 10^{19} GeV, it's true, but getting uncomfortably close.

And besides, by the late 1970s, cosmologically minded physicists were starting to turn up some serious problems with the extrapolation. It wasn't just that they were (and are) unclear on which specific models of grand unification stood a chance of being correct—if any were. It's that what they *did* know seemed to be at odds with reality.

In 1978, most notably, Soviet theorists had pointed out that a GUT theory's symmetry-breaking fields wouldn't have crystallized out of the cooling primordial fireball in a nice, uniform manner. Instead, these fields would have accumulated a hodge-podge of "crystal defects"—flaws that would have been something like the cracks and imperfections seen in diamonds, ice cubes and other earthbound crystals. Furthermore, many of these flaws would have evolved into magnetic monopoles—hypothetical particles that were what you would get if you could somehow isolate the north and south poles of a bar magnet and move them far apart, so that they looked like the magnetic analog of isolated electric charges [28] (Fig. 4.4).

Unfortunately, there were two major problems with this prediction. One was what every student learns in physics 101: Magnetic monopoles don't exist. Break a real bar magnet, and you just get two short bar magnets, each with its own north and south pole. There is no way to isolate them.

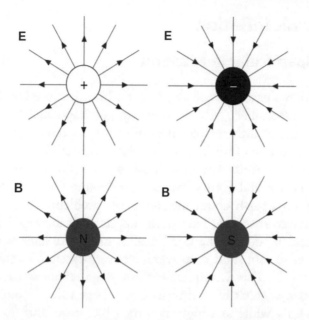

Fig. 4.4 Top: Isolated electric charges such as electrons, protons, atomic nuclei can be thought of as a place where electric field lines begin or end. **Bottom**: In the 1970s, physicists realized that their grand unified theories of particle physics predicted that the early universe would have produced huge numbers of *magnetic monopoles*—in effect, isolated north or south magnetic poles that would be the end- or starting-points of magnetic fields (here labeled **B**). But no such magnetic monopoles have ever been observed. The challenge for cosmologists was to figure out why not (Maschen/Wikimedia Commons, CC0)

Of course, it was always possible that magnetic monopoles did exist somewhere in the universe, and scientists had just never noticed them. But that was the second problem: The calculations also showed that these things would have a mass on the order of the GUT scale, 10^{16} GeV, or about 10 million billion times the mass of a proton. And there would have been *lots* of them, making them really hard to miss [29].

Indeed, the GUT transition in general and the monopole problem in particular left some of the best theorists in the world scratching their heads—until the closing days of the decade, when one young physicist had a spectacular idea…

4.2   Cosmic Inflation

4.2.1   A Spectacular Realization

Alan Guth knows exactly when the idea hit him, because it's right there in his notebook: December 7, 1979—a Friday, as it happened.

Guth had just recently arrived in California for a temporary postdoctoral appointment at the Stanford Linear Accelerator Center, after a stint at Cornell University in New York. And cosmic phase transitions were much on his mind [30, 31]. He and Henry Tye, another young Cornell physicist, were just finishing up a paper about the monopole problem. Their idea was that monopole production might have been suppressed if the expanding universe had "super-cooled" during the GUT era, in much the same way that water can sometimes remain fluid even when it's (very carefully) cooled below its freezing point [32]. This was equivalent to saying that one of the GUT scalars might have gotten stuck for a time at a very large value,[16] like rainfall that gets trapped for a while in a high mountain lake instead of flowing straight down to the sea. (The known GUT theories left a lot of ambiguity in the equations that govern these fields, so why not?).

This super-cooling scenario worked, sort of. But on that December Friday, Tye was getting ready for an extended visit to his native China, and had suggested that Guth use the time to give the whole thing a rethink. In the earlier paper, the two of them had just assumed that the universe would continue to expand in the same way no matter what the GUT scalars were doing. But was that really true?

Not at all. When you combined any of the GUT theories with the Friedmann-Lemaître cosmological equations, Guth realized on that December Friday, the term that described this trapped field would look just like a cosmological constant.

A really, really *big* cosmological constant.

The resulting outward force wouldn't produce the nice, calm, static universe that Einstein had imagined, balanced right on the knife edge between expansion and contraction. With *this* cosmological constant, Guth

---

[16] Technically speaking, the GUT equations said that scalar field would have both a *kinetic energy* that depended on its rate of change, and a *potential energy* that depended on the value of the field itself. The symmetries of any given GUT theory don't actually say what this potential energy function is, unfortunately, which leaves a lot of room for the kind of theorizing that Guth and Tye were doing—as well as the theorists we'll meet later in this chapter. But the resulting math is very similar to the equations that describe a ball (the field) rolling around in an undulating landscape (the potential energy), a situation familiar to any Physics 101 student. That's why researchers in this field almost always use the landscape metaphor.

saw, the universe would have exploded outwards at rates that would have dwarfed the already mind-boggling speeds predicted in standard cosmologies. The universe would experience an *exponential* expansion, doubling its size in an infinitesimal slice of time—then doubling its size again, and again, and again, at least 90–100 times. In considerably less than a nano-nano-nanosecond, such an exponential expansion could have inflated the universe by a factor of 10^{28} or more. To put that factor in perspective, it was like growing today's observable universe, now roughly 93 billion light-years wide, from a patch of space the size of a softball.

So much for the monopole problem. No matter how tightly the things were packed when they formed, Guth reasoned, this cosmic inflation would have spread them so far apart that there would be maybe one left in our entire observable universe. We'd never notice it.

The Flatness Problem

But there was more. The previous year, Guth had heard Princeton University physicist Robert Dicke give a lecture about the flatness problem: a cosmological conundrum that almost no one but Dicke seemed to worry about.

Dicke had found this puzzle hidden inside the Friedmann-Lemaître cosmological equations—the part where the mathematics tells us that the overall shape of the universe is largely determined by the matter and energy it contains (Fig. 4.5). There are three possibilities. If the density of matter and energy is higher than a certain critical value, then the Big Bang will produce a cosmic sphere that expands, slows to a halt, and then comes crashing back down into a Big Crunch. It's a bit like throwing a baseball upward and watching it fall back to Earth.

If the density is lower than the critical value, however, then the universe will emerge from the Big Bang with infinite extent but negative curvature—a *hyperbolic* space that's like a mountain pass, or a saddle that curves upward in one direction and downward in the other. Such a universe will expand forever, like launching the baseball into space with a rocket that can exceed Earth's escape velocity (about 11.2 km per second), and watching it sail off to infinity.

Finally, though, if the cosmic density is right at the critical value— meaning, right on the boundary between too high and too low—then the universe will still be infinite in extent and will still expand forever. But mathematically speaking, it will be as flat as an endless sheet of paper. This is like launching the baseball right *at* escape velocity: It will still keep going forever, but its speed away from Earth will asymptotically fall toward zero.

And therein lay the flatness problem, said Dicke. If the cosmic density had been even slightly too high back in the GUT epoch, then the closed universe would have experienced more of a Big Burp than a Big Bang: It would have expanded and then contracted again within about a Planck time. Conversely, if the cosmic density had been even slightly too low, the open, infinite universe would have expanded too fast for stars or galaxies to form. Yet here we are, more than 10 billion years later, with stars and galaxies everywhere and an average matter density that's no more than a factor of ten away from the critical density (equivalent to a few hydrogen atoms per cubic meter.) For that to be true today, Dicke argued, then the cosmic density during the GUT epoch would have had to match the critical density to one part in 10^{55}—a precision that made balancing a pencil on its point look trivial.

It was also a precision that demanded explanation. The standard cosmological models offered none. They just took today's cosmic density as a given: It is what it is. But that, as Guth realized with a growing excitement, was because the standard cosmologies didn't take inflation into account. Once you did, the flatness problem went away as easily as the monopole problem had: No matter how the universe started out, inflation's exponential expansion would stretch it to be as taut and as flat as the surface of a balloon that's been blown up to the scale of light years or more.

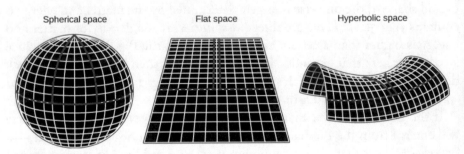

Fig. 4.5 According to Einstein's general theory of relativity, the overall shape of the universe is determined by the density of matter and energy. **Left**: If the average density is higher than a certain critical value (equivalent to a few hydrogen atoms per cubic meter) the universe will be an enormous, but finite sphere, and its expansion will eventually reach a maximum size and reverse. Such a universe will end in a Big Crunch—essentially, a Big Bang in reverse. **Right**: If the average density is less than the critical value, the universe will be an infinite hyperbolic space shaped like a mountain pass, or saddle, and will expand forever. **Center**: If the density is right *at* the critical value, the universe will be shaped like flat, infinite plain and will still expand forever—but will do so at a rate that falls toward zero over time. Our universe appears to be very close to flat (Image credit: (Franknoi A, Morrison D, Wolff SC (2016) Astronomy, Houston, Texas. Access for free at https://openstax.org/books/astronomy/pages/1-introduction, CC BY 4.0)

Guth was so thrilled with this insight that he wrote it down in his notebook that evening and drew a double box around the whole paragraph:

SPECTACULAR REALIZATION:

This kind of supercooling can explain why the universe today is so incredibly flat – and therefore resolve the fine-tuning paradox pointed out by Bob Dicke in his Einstein Day lectures.

The Horizon Problem

Spectacular though this realization already was, however, there was still more. Just a few weeks later, Guth realized that inflation could also answer a second cosmological conundrum: Why, on average, does the universe look the same in every direction? Why do astronomers find the same statistical distribution of galaxies in the northern sky as in the south? And why is the temperature of the CMB the same everywhere in the sky?

Again, most cosmological models just assumed this cosmic uniformity, building it into the equations without trying to explain it: The universe is what it is. But, as the Austrian-born physicist Wolfgang Rindler and others had been pointing out since the 1950s, this degree of regularity is actually very strange [33–35]. The early universe wasn't like soup that's simmering on a stovetop, where there is plenty of time for heat to flow, for ingredients to blend, and for temperature and taste to equalize throughout the pot. The early universe was expanding so fast that particles generally couldn't make it from one region to another, even if they were photons moving at the speed of light. (Imagine an ant crawling on the surface of a balloon that's getting blown up faster than it can move: Its destination just keeps getting further and further away.)

As a result, the primordial photons coming at us from different parts of the sky are streaming in from regions of the distant, early universe that had no time to communicate with one another, much less time to mix and come to a common temperature. Nor, for that matter, would these regions have been able to synchronize their expansion, so that every part of the universe started at the same time. Unless you had two patches of cosmic plasma that were unimaginably close to start with, the subsequent history of one would have been completely independent of the other.

And yet these patches all ended up in sync, looking the same in every part of the sky. How?

In technical terms, any region of the universe that can't get signals to you, even at the speed of light, is said to be beyond your *horizon*. So this became known as the horizon problem. And as soon as Guth learned about it at a lecture he attended in January 1980, he realized that inflation would turn *this* into a non-problem, as well. Thanks to that 10^{28+} exponential expansion factor, everything that would grow into today's observable universe *would* have started out unimaginably close, packed into a space considerably smaller than a proton. Or to turn that around, inflation would have taken an infinitesimal patch of the universe so small that it *did* have time to mix and become uniform, and expanded it to cosmic dimensions. Everything you can see today, from your morning coffee to the microwave background, was once a part of that primordial mix (Fig. 4.6).

A Graceful Exit?

Guth considered this solution of the horizon problem to be yet another compelling argument for inflation. Yet he was also acutely aware that inflation had problems of its own—the biggest being the mystery of its ending.

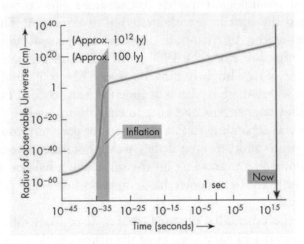

Fig. 4.6 Although the quantum era that prevailed before the Planck time at $t = 10^{-43}$ s is a mystery, the universe expanded out of that era with its temperature falling precipitously. And as a result, the grand unification that encompassed all the known forces (except gravity) began "crystallize" at about $t = 10^{-35}$ s, breaking apart into the strong, electromagnetic, and weak forces we know today. But most cosmologists believe that this breakup triggered a period of exponential expansion known as cosmic *inflation*. Within far less than a nanosecond, the universe would have inflated by a huge factor—10^{40} in this diagram—which would have been enough to eliminate three cosmological mysteries known as the monopole, flatness, and horizon problems (*Image credit*: Mughal et al. [114])

It obviously *did* end, since the universe is not inflating now.[17] But everything depended on how, exactly, that happened.

Guth's initial assumption was that inflation would stop once quantum effects allowed the GUT scalars to tunnel their way out of the high mountain valley that trapped them. Once these fields reached the sea, so to speak, they would no longer look like a huge cosmological constant, inflation would cease, and the cosmic expansion would slow down to the merely mind-boggling rates called for in the standard Friedmann-Lemaître equations.

What made this picture appealing was that the energy released by these tunneling fields would produce hordes of new particles, thereby resolving yet another cosmic mystery: Where did all those quarks, photons, electrons, and such come from in the first place? The answer was that they came from the energy that powered inflation. In fact, that's what the Big Bang *was* in this picture: The end of inflation.

To Guth's frustration, however, his detailed calculations of this tunneling scenario revealed it to be a train wreck. Instead of producing the statistically smooth, flat distribution of matter that we see around us, the tunneling would have ended up producing isolated bubbles of normal space that would never merge, and that would be grossly inhomogeneous. Or to put it another way, the exit from inflation would have destroyed everything inflation achieved.

Still, inflation solved so many problems that Guth felt there had to be something *right* about it. So when he finally published his paper in January 1981, a little over a year after his spectacular realization, he frankly admitted in the text that he couldn't solve the "graceful exit" problem. Instead, he appealed to his fellow physicists for help: "I am publishing this paper in the hope that it will … encourage others to find some way to avoid the undesirable features of the inflationary scenario" [36].

4.2.2 New Inflation—With Lumps

He didn't have long to wait. Guth's inflation idea soon drew the attention of Andrei Linde at the Lebedev Physics Institute in Moscow, where (unknown to Guth) Linde's colleague Alexei Starobinsky had recently suggested a somewhat similar inflationary scenario [37]. The early universe was already familiar terrain to Linde, who had been one of the first to recognize that unified

---

[17] Or at least, it's not inflating at that same prodigious pace; as we will see in the next chapter, today's universe seems to be undergoing a kind of micro-inflation that's slowly increasing the cosmic expansion rate.

theories would undergo symmetry-restoring phase transitions [16], and that a Higgs-like field might look like a cosmological constant [38].

And now Linde was quick to come up with an alternative graceful-exit scenario, dubbed *slow-roll* inflation. Instead of asking the GUT scalar to tunnel out of a high mountain valley, so to speak, he imagined it perched atop the mathematical equivalent of a broad, flat dome similar to Ayers Rock in Australia, or Stone Mountain in Georgia. (This shape was known in the trade as a Coleman-Weinberg potential [39], and had been studied by others in non-inflationary contexts.) Like an ordinary marble placed on such a dome, Linde argued, the field would roll down very slowly at first—with the universe inflating furiously every instant that it was up there. But when it finally did plunge toward the bottom and convert its energy into swarms of particles, he showed, the resulting bubble of normal space would be more than big enough to accommodate our entire observable universe. No more gross inhomogeneity (Fig. 4.7).

Linde published this result in 1982—and so did American physicists Andreas Albrecht and Paul Steinhardt, who had independently arrived at the same idea [40–42]. Their "new inflation," as it came to be known, put the idea on a much more solid footing—and, not incidentally, gave astronomers a way that they could look at the sky and actually test all this theorizing.

The test started from the observation that (new) inflation was *too* successful. By itself, it would have produced a cosmos stretched so flat that there would be nothing in it but a uniform haze of hydrogen and helium gas, with no discernable lumps. So where did the universe get all those lumps that we now call galaxies and clusters?

The answer again went back to Sakharov, who had pointed out in 1966 that quantum fluctuations during the very first instants of the Big Bang would have caused subtle variations in the density of the primordial plasma [43]. As small as these variations were to begin with, Sakharov noted, they would have expanded along with the rest of the universe. And over the billions of years since, the inexorable pull of gravity would have caused the denser regions to contract and grow denser still, until they condensed into galaxies, clusters, and other massive structures—among them our own Milky Way galaxy.

If Sakharov was right, in other words, we owe our existence to quantum fluctuations that occurred billions of years ago, in the universe's first infinitesimal fraction of a second.

Physicists at the Lebedev Institute revived Sakharov's idea in 1981, and applied it to the version of inflation developed by their colleague Starobinsky [44, 45]. Then a year later, other teams of physicists rushed to do the same for the recently invented theory of new inflation; they published their work

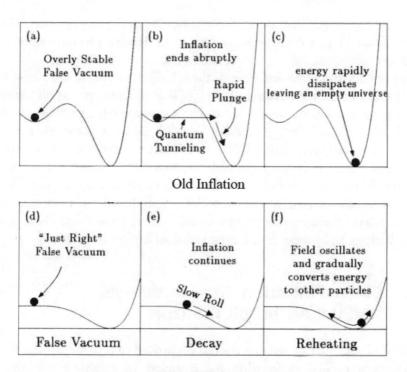

Fig. 4.7 The math that governs a grand unified theory's symmetry-breaking scalar field(s) is very similar to the elementary physics equations for a ball (the magnitude of field) rolling around in an undulating landscape (the field's potential energy). Unfortunately, the various GUTs are vague about what this landscape actually looks like. **a–c** Guth's original guess predicted an exit from inflation that was too abrupt, and would not have produced a universe like the one we live in. **d–f** An alternative "slow-roll" exit from inflation, which assumed a slightly different landscape, offered much more realistic-looking results (*Credit* Mughal MZ, Ahmad I, García Guirao JL (2021) Relativistic Cosmology with an Introduction to Inflation. Universe 7:276. https://doi.org/10.3390/universe7080276, CC BY 4.0)

in four separate papers that grew out of the Nuffield Symposium: a two-week workshop organized by Stephen Hawking and others in Cambridge, UK, during the summer of 1982 [46–49].

The consensus in these papers was that the GUT scalar field—or, as many physicists took to calling it, the *inflaton*—would indeed undergo quantum fluctuations while it was slowly rolling off that high dome. Intuitively speaking, it was as if the inflaton were a marble being randomly jostled as it moved. At some points in the infant universe the marble would randomly be kicked forward, so to speak, so that the graceful exit from inflation got accelerated ever so slightly. But at other points the marble would randomly be kicked backwards, and the exit would be ever so slightly retarded. The net effect would be just the kind of variations in density and temperature that Sakharov had predicted.

Better still, the physicists agreed, these variations might just show up as subtle shifts in the temperature of the CMB at different points on the sky.

They were right. As we discuss in the next chapter, researchers working with NASA's Cosmic Background Explorer (COBE) satellite would announce in 1992 that they had found CMB temperature variations with exactly the properties expected from quantum fluctuations [50]. Moreover, these findings would continue to hold up when later satellites mapped the CMB in much finer detail. And in the meantime, a multitude of computer simulations would confirm that gravity could indeed have taken these density perturbations and produced the observed distribution of galaxies.

4.3   Eternal Inflation, the Multiverse, and the Anthropic Principle

In sum, Alan Guth's spectacular realization proved to be a spectacular success, providing a unified explanation for a variety of otherwise disconnected observations [51].

- Cosmic inflation solved the flatness problem, and thereby explained why the universe is both big and old.
- It solved the horizon problem, and thereby explained why the universe is so uniform on average.
- In its revised version, it solved the "graceful exit" problem, and thereby explained where matter comes from. And…
- It solved what might be called the lumpiness problem, and thereby explained how the universe acquired the subtle variations in density that ultimately gave rise to galaxies, stars, planets, and us.

All of which is why cosmic inflation quickly became widely (if not quite universally) accepted, just as it is today.

Swiss-Cheese Inflation—And Metaphysical Cosmology

But again, theorists didn't stop there. At the same Nuffield Symposium where he and others calculated the quantum fluctuations produced during new

inflation, Paul Steinhardt pointed out some of the theory's more radical implications [52]. Just for fun, he said, let's imagine that you have a cosmos that's
inflating away, with an inflaton field slowly rolling down from its mountaintop. If you look at this cosmos on a local scale—"local" meaning the size
of our observable universe—then you've just got the standard new-inflation
scenario, with the energy of the inflaton turning into a firestorm of normal
particles as inflation ends and the considerably slower Big Bang starts up.

But now, said Steinhardt, imagine that you step back and look at this (infinite) cosmos on a really *big* scale—say, 10^{30} times the size of our observable
universe. Then you might find that what we naïvely call *the* universe is just
an isolated bubble of normal space surrounded by an endless expanse of space
that's still inflating. After all, there's nothing that says the inflaton field had
to roll down in lockstep everywhere; our local cosmic bubble could just be a
region where it got to the bottom earlier than some—and later than others.
In fact, noted Steinhardt, our über-cosmos could be as full of normal-space
bubbles as a fractal Swiss cheese, even as inflation kept opening up room for
more.

But wouldn't we have noticed such a thing? Not necessarily, said Steinhardt. The walls of any given cosmic bubble would expand at essentially the
speed of light, so that observers who are deep in the interior—as we are—
would never see them. Nor were we likely to see another bubble universe
colliding with ours: inflation in the larger cosmos would drive the bubbles
apart much faster than light.[18]

In fact, this scenario would have no observational consequences at all—
which is why Steinhardt whimsically gave this whole field of inquiry the name
"metaphysical cosmology."

Meta or not, however, the idea was so compelling that others quickly
picked up on it. At Tufts University in Massachusetts, for example, Alexander
Vilenkin showed that the fractal Swiss-cheese picture held true not just for the
specific version of slow-roll inflation that Steinhardt had considered, but for
any form of new inflation [53].

Eternal, Chaotic Inflation?

Linde, meanwhile, was pointing out that you didn't even need new inflation—the slow-roll theory he'd just helped invent. You'd still get inflation if
the marble were simply rolling around inside a bowl, so to speak—a model

---

[18] Einstein's relativistic speed limit applies only to individual particles. Space, as it turns out, can
expand as fast as it wants to.

of the inflaton field that was even more natural and plausible than the slow-roll mountaintop version. Once you took careful account of how cosmic expansion influences any quantum field, Linde explained, you'd find that it produces a "friction" term in the equations, as if the sides of the bowl were coated with thick molasses. So if you had a region of the cosmos where the inflaton field just happened to be large, he said, then the metaphorical marble would be stuck way up on the sides, and that region would inflate like mad during the nano-nano-nano-nanoseconds it took for the marble to ooze down to the bottom and give rise to a normal Big Bang. And if you likewise assumed that the inflaton field was distributed randomly, said Linde, then the patches of the cosmos where the field just happened to be largest would inflate the fastest.

The result—once you also took account of the inflaton's quantum fluctuations—would be a cosmos of interlocking bubbles that looked less like Swiss cheese and more like an ever-expanding, fractal Christmas tree. Linde called this "the eternally existing, self-reproducing, chaotic inflationary universe" [54–56].

Eternal? Yes, said Linde– because in a cosmos that's infinite in spatial extent, there's no reason that this web of endlessly proliferating bubble universes couldn't also extend infinitely far into the past and future. Each bubble would have its own beginning of time: the point where its local inflation ceased and gave way to a standard Big Bang. But the whole "multiverse" (to use the modern term) might very well have no beginning, no end—and no need for a $t = 0$ singularity. In fact, it would exist as a kind of steady-state cosmos—albeit in a way that neither Arthur Eddington nor Fred Hoyle had ever imagined.

The Anthropic Principle

This (possible) elimination of the singularity problem was one of the two big things about Steinhardt's scenario that had fascinated Linde from the start [54, 57]. The other was the infinite abundance of new bubble universes. What if these other creations weren't just physically separated from ours, Linde asked himself, but followed separate paths as the Higgs field and all its relatives crystallized? What if these bubbles wound up with different patterns of symmetry breaking, different particle masses, different values for the electric charge and all the other coupling constants—maybe even a different number of dimensions. What if they had different physical laws entirely?

That question then led to another: What if only a few of the bubble universes had laws conducive to life? If that were the case, reasoned Linde, then our universe must obviously belong to that very small set. And that, in turn, might go a long way toward explaining some of the odd coincidences in cosmology (Fig. 4.8).

Linde was hardly the first to wonder about these coincidences. As far back as 1917, for example, the Austrian-Dutch physicist Paul Ehrenfest had been thinking along the same lines when he asked why our universe has only three spatial dimensions (plus the one dimension of time). After all, Ehrenfest noted, it's mathematically possible to have any number of dimensions [58]. His answer, backed up by detailed calculations, was that Maxwell's theory of electromagnetism wouldn't work in anything other than three spatial dimensions—nor would a planet's orbit around the sun be stable. So in anything other than the universe we have, Ehrenfest argued, there would be no atoms, no light, no solar systems, and presumably, no life.

In 1961, Princeton's Robert Dicke had used similar reasoning when he noted that physicists don't get just any old number when they measure the age of the universe [59]. After all, he quipped, "it is well known that carbon is required to make physicists." So before there can be anyone around to make observations, there has to be enough time for galaxies to form, for stars to evolve, and for nuclear reactions in those stars to produce carbon and all the

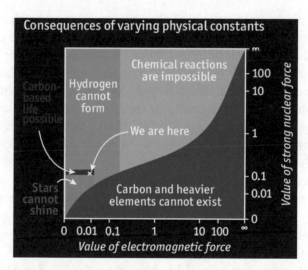

Fig. 4.8 The *anthropic principle* says that the constants of nature have the values they do because otherwise, no one would be here to observe them. This chart gives an example using just two constants: the strengths of the electromagnetic and strong forces. In theories that imagine a near infinity of universes with different physical laws, like eternal inflation or the superstring landscape, the anthropic principle is simply the statement that only a very small fraction of those universes would be compatible with life (*Credit* Dąbrowski MP (2019) Anthropic selection of physical constants, quantum entanglement, and the multiverse falsifiability. Universe 5:172. https://doi.org/10.3390/universe5070172, CC BY 4.0)

other heavy elements. Since all this takes billions of years, Dicke noted, it's no surprise that observers find a cosmic age of roughly that magnitude (The modern number is 13.8 billion years.)

In 1974, the Australian physicist Brandon Carter had given this deceptively simple idea a name, the *anthropic principle*, and had provided many more examples [60]. One was the constant G that appears in Newton's law of gravity, as well as in Einstein's general relativity. This constant determines how strong the gravitational force is, but neither Newton nor Einstein could explain why G has the (very small) value that's observed. It seemed to be an arbitrary number that had to be determined by experiment. But Carter pointed out that if we lived in a universe where G were very much smaller or larger than it is, then planets could not have coalesced from interstellar gas, and we wouldn't be here to worry about it.

Likewise with the strong force that holds protons and neutrons together in the nucleus. If it had been even a little bit weaker than it is, Carter noted, then helium wouldn't have formed in the early universe, and hydrogen would have been the only element. If the force had been a little stronger, though, then the universe would have become nothing *but* helium. Either way, no planets.

Other physicists soon suggested that the anthropic principle could also account for the mass and electric charge of the electron, the cosmological constant, and any number of other physical parameters [61–65].

But the anthropic principle coupled with the inflationary multiverse idea did even more: It gave physicists a plausible reason to imagine that *our* universe might not be *the* universe. Instead, we might be just one experiment carried out by a cosmos that is forever trying out an infinity of different (bubble) universes with infinite variations in physical law. And that prospect, in turn, made it much more plausible that the laws we see around us are a selection effect.

To Linde and the handful of others who continued to study this idea through the 1980 and 1990s, this was a period of exciting advances—including new insights into how inflation could have given rise to baryogenesis [66]. But the pursuit was also a lonely one for much of that time, he later admitted, even after he and his wife, the physicist Renata Kallosh, left Russia for Stanford University in 1990. To most researchers, the notion of unobservable other universes sounded like just so much metaphysics, not to mention comic-book science fiction.

That attitude started to change in 1998, however, when two teams of astronomers announced that the expansion of the universe is actually accelerating [67, 68]. We'll be giving a fuller discussion of this discovery in the next chapter. But the key point for now is that the "dark energy" driving this acceleration looks just like Einstein's cosmological constant, which presented theorists with a conundrum. Before, they had been struggling to understand why every attempt to measure the cosmological constant seemed to yield zero, while most attempts to calculate it in quantum field theories gave values that were enormous.[19] Now their challenge was even tougher: they had to explain a cosmological constant that wasn't quite zero—a number that would at least have been consistent with theories of supersymmetry—yet was smaller than the field-theory estimates by a factor of roughly 10^{-120} [51, 66].

The difficulty of this challenge not only pushed Einstein's mysterious constant to the forefront of theoretical physics, but began to revive interest in earlier anthropic explanations for its infinitesimal size [63–65, 69].

The Superstring Landscape

Meanwhile, the anthropic principle was getting an additional boost from superstring theory: a mathematical framework that had become very popular in the 1980 and 1990s, and that is still widely considered to be about as close as anyone has ever come to a fully unified account of particle physics and gravity [70]. This theory is built around the notion that particles are actually infinitesimal threads of energy—the superstrings—and that these threads have to move around and vibrate in 10 dimensions: 9 directions of space plus one of time. Indeed, this number 10 is actually a prediction of the theory: Try to push the number of dimensions lower or higher, and the equations become mathematically inconsistent. So, since we obviously live in a universe with just three space dimensions, the others must be curled up so tightly that we can't see them. (Think of rolling a 2-D sheet of paper into a thin straw; from a distance, the straw looks like a 1-D line.)

String theorists had always known that there are any number of ways to carry out such a "compactification," each yielding a different pattern of symmetry breaking, a different dimensionality, and all the rest. But it was only in the early 2000s that they began to understand how this might play out in a cosmological context—and just how *many* compactifications there could be [71–74]. Researchers in the field were soon talking about an undulating string-theory "landscape" with something like 10^{500} valleys that corresponded to stable compactifications.

---

[19] Quantum field theories predict that "empty" space is actually filled with huge amounts of cosmological-constant-like quantum energy. So if the observed cosmological constant is very, very small, this vacuum energy has to be cancelled by something. But what?

That's a number so large as to defy all metaphor. And, since there was no obvious reason for nature to prefer one compactification over another, it's a number that did a lot to boost the anthropic principle's reputation. It wasn't hand-wavy anymore: Physicists now had a rigorous example of how one underlying theory—superstrings—could produce an inconceivable multitude of universes with different-seeming physical laws. And that made it much more plausible that the particular universe we see around us, including our infinitesimally-small-but-not-zero cosmological constant, was the result of anthropic selection [75].

In short, as Guth described in a 2007 review of inflation's first quarter-century [51], the confluence of dark energy, the string theory landscape, and the eternally inflating multiverse made anthropic reasoning downright respectable. A decade later, Linde's own review [66] described how the resulting influx of researchers "transformed this field into a vibrant and rapidly developing branch of theoretical physics."

4.4   Alternatives to the Multiverse?

It must be said, however, that many other researchers consider this whole line of reasoning to be anathema. If nothing else, they see it as intellectually lazy: Anything you don't understand, just wave your hand and say, "anthropic" [76, 77]. Guth, now a senior professor at MIT and himself a cautious believer in the multiverse, wrote in his 2007 review that "anthropic reasoning means the end of the hope that precise and unique predictions can be made on the basis of logical deduction"—a hope that is not to be given up lightly. Or, as Steinhardt put it as he was explaining why he has rejected his own multiverse idea, "a theory that predicts everything predicts nothing" [78].

That said, however, rejecting the anthropic principle means rejecting infla-tion, since the latter will almost inevitably give rise to a multiverse. This has led some cosmologists to propose ways of testing the multiverse idea obser-vationally [79–81].[20] But it has also led others to explore ways of solving the flatness problem, the horizon problem, and all the rest *without* inflation.

Variable Constants?

Several investigators have suggested that these problems would solve them-selves if the fundamental constants of physics weren't really constants—that

---

[20] The thinking is that early collisions with other bubble universes might have produced subtle pertur-bations in our universe's microwave background, and that these perturbations might, in principle, be detectable.

is, if parameters such as the gravitational constant, Planck's constant, and even the speed of light took on very different values during the earliest instants of the universe.

The idea of time-varying constants is not a new one. The British physicist Paul A. M. Dirac, one of the founders of quantum mechanics, raised the possibility as early as 1937 [82, 83], and Dicke explored the notion further in 1961 [59]. But no one saw the idea as an alternative to inflation until 1993, when University of Toronto physicist John Moffat realized that the horizon problem would go away if the speed of light had been much, much faster in the early universe [84, 85]. The higher speed limit would have allowed the cosmic patch that is now our observable universe to have mixed completely before it expanded, thereby ensuring that today's universe would still look the same on average in every direction. A similar analysis allowed Moffat to conclude that a dramatically faster speed of light would also make the flatness and monopole problems disappear.

In 1999, Albrecht and João Magueijo independently rediscovered this idea and reached much the same conclusions [86]. Magueijo, a Portuguese-born physicists working at Imperial College London, would go on to become one of the most ardent champions of the variable-constant approach [87]. In 2004, for example, he and Lee Smolin introduced "rainbow gravity": a variant of general relativity in which different frequencies, or colors of light experience a slightly different level of gravity, causing them to split and take different paths [88]. This effect would not be detectable on Earth, but could become important in regions where the curvature of space–time is extremely high, such as near a black hole. Magueijo and Smolin argued that rainbow gravity could solve the horizon problem. And in 2013, a group of Egyptian and Indian physicists showed that such models could also do away with the initial cosmic singularity; they found that if you use this framework to trace the path of matter and light back 13.8 billion years, not all the light rays would trace back to the same place at the same time. So there would be no infinitely small point of origin; instead, the cosmos would have an infinitely long tail as you went backwards, shrinking toward zero at an exponentially slower rate, but never reaching it [89].

In more recent work, Magueijo and his collaborators have explored other possibilities. Perhaps gravity didn't even exist in the early universe, for example, and turned on only after the expanding cosmos had cooled to a certain point. That would eliminate the need to reconcile quantum theory and gravity, and thereby get rid of a one of the biggest headaches in theoretical physics [90]. Or perhaps gravity did exist in the early universe, but moved at a speed different from light [91]. Making this second assumption mathematically consistent requires that the speed of light vary over time in a particular way, and predicts primordial density fluctuations that turn out to be very close to the ones we observe—without the need for inflation [92].

Loops and Cycles?

Among physicists hoping to unify gravity with quantum physics, easily the most popular approach is the superstring theory mentioned earlier. But a strong runner-up is *Loop Quantum Gravity* (LQG), a framework proposed in the 1980s by Abhay Ashtekar [93], and further developed in the 1990s by Ashtekar, Smolin, Carlo Rovelli and others [94–98]. The LQG model, posits that space emerges from a network of tiny geometrical loops (not to be confused with superstrings). It also imposes a fundamental minimum size limit, suggesting that the universe could never have been squashed down into an infinitely small point at the Big Bang. Indeed, this suggestion was made explicit in the early 2000s with the development of Loop Quantum Cosmology: a family of approximations to LQG that are roughly analogous to the classical cosmological equations that Friedmann and Lemaitre derived from Einstein's general relativity [99–105]. As the name suggests, Loop Quantum Cosmology allows for a fully quantum mechanical analysis of the early universe, and leads to a picture in which the universe was once large and contracting, and shrank down to a minimum size around 13.8 billion years ago before growing again. There was thus no Big Bang in this picture, but a Big Bounce—and we live in the expanding post-bounce phase [106]. In fact, some versions of the model update the *cyclic* cosmology idea once considered by Einstein himself, not to mention Robert Dicke, and predict that the universe cycles through a series of such bounces, expanding and then contracting repeatedly. Either way, this picture does not invoke an inflationary multiverse—but it is compatible with an inflationary phase occurring post-bounce, with predictions that fit with the observed patterns in the CMB [107].

Steinhardt, meanwhile, has co-authored at least two inflation alternatives. One is the *ekpyrotic* theory, in which our 4-dimensional spacetime is supposed to be a membrane embedded in some higher-dimensional cosmos, and the Big Bang resulted from a collision with another, parallel membrane [108]. The second alternative is yet another version of cyclic cosmology idea: Steinhardt and his co-workers propose that the universe is smoothed to the requisite flatness and homogeneity during the contraction phase, before bouncing back outwards again [109].

It's not clear how viable such models are, though; Linde, not surprisingly, is an extreme skeptic of cyclic cosmology, and has marshalled many arguments against it [110, 111]. Certainly it's fair to say that no version of the idea has achieved anything like the widespread acceptance enjoyed by inflationary cosmologies.

But then, there's not likely to be any final resolution of these debates until physicists manage to come up with that long-sought merger of quantum theory and general theory—a merger that, whether it's superstrings, quantum loops or something else entirely, is widely expected to encompass not just the origin of the universe, but the origin of space and time themselves.

References

1. Weinberg S (2015) Steven Weinberg—Session I. In AIP oral histories. https://www.aip.org/history-programs/niels-bohr-library/oral-histories/339 96-1. Accessed 22 May 2020
2. Oerter R (2006) The theory of almost everything: the standard model, the unsung triumph of modern physics, Reprint Edition. Plume
3. Planck M (1901) Ueber das Gesetz der Energieverteilung im Normalspectrum. Ann Phys 309:553–563. https://doi.org/10.1002/andp.19013090310
4. Nauenberg M (2016) Max Planck and the birth of the quantum hypothesis. Am J Phys 84:709–720. https://doi.org/10.1119/1.4955146
5. Kolb E, Turner M (1994) The early universe. CRC Press, Boulder, Colo
6. Steigman G (2006) Primordial nucleosynthesis: successes and challenges. Int J Mod Phys E 15:1–35. https://doi.org/10.1142/S0218301306004028
7. Fritzsch H, Gell-Mann M, Leutwyler H (1973) Advantages of the color octet gluon picture. Phys Lett B 47:365–368. https://doi.org/10.1016/0370-269 3(73)90625-4
8. Rafelski J (2015) Melting hadrons, boiling quarks. Eur Phys J A 51:114. https://doi.org/10.1140/epja/i2015-15114-0
9. Yang CN, Mills RL (1954) Conservation of Isotopic Spin and Isotopic Gauge Invariance. Phys Rev 96:191–195. https://doi.org/10.1103/PhysRev.96.191
10. Glashow SL (1961) Partial-symmetries of weak interactions. Nucl Phys 22:579–588. https://doi.org/10.1016/0029-5582(61)90469-2
11. Weinberg S (1967) A model of Leptons. Phys Rev Lett 19:1264–1266. https://doi.org/10.1103/PhysRevLett.19.1264
12. Salam A (1968) Weak and electromagnetic Interactions. In: Elementary particle theory. Relativistic groups and analyticity. Wiley, pp 367–377
13. The Nobel Prize in Physics (1984) NobelPrize.org. https://www.nobelprize.org/prizes/physics/1984/summary/. Accessed 24 Dec 2021

14. Chatrchyan S, Khachatryan V, Sirunyan AM et al (2012) Observation of a new boson at a mass of 125 GeV with the CMS experiment at the LHC. Phys Lett B 716:30–61. https://doi.org/10.1016/j.physletb.2012.08.021

15. Aad G, Abajyan T, Abbott B et al (2012) Observation of a new particle in the search for the Standard Model Higgs boson with the ATLAS detector at the LHC. Phys Lett B 716:1–29

16. Kirzhnits DA, Linde AD (1972) Macroscopic consequences of the Weinberg model. Phys Lett B 42:471–474. https://doi.org/10.1016/0370-2693(72)90109-8

17. Weinberg S (1974) Gauge and global symmetries at high temperature. Phys Rev D 9:3357–3378. https://doi.org/10.1103/PhysRevD.9.3357

18. Dolan L, Jackiw R (1974) Symmetry behavior at finite temperature. Phys Rev D 9:3320–3341. https://doi.org/10.1103/PhysRevD.9.3320

19. Sakharov AD (1967) Violation of CP Invariance, C Asymmetry, and Baryon Asymmetry of the Universe. Sov J Exp Theor Phys Lett 5:24

20. Weinberg S (1979) Cosmological production of baryons. Phys Rev Lett 42:850–853. https://doi.org/10.1103/PhysRevLett.42.850

21. Ellis J, Gaillard MK, Nanopoulos DV (1979) On the effective lagrangian for baryon decay. Phys Lett B 88:320–324. https://doi.org/10.1016/0370-2693(79)90477-5

22. Abe K, Akutsu R, Ali A et al (2020) Constraint on the matter–antimatter symmetry-violating phase in neutrino oscillations. Nature 580:339–344. https://doi.org/10.1038/s41586-020-2177-0

23. Pati JC, Salam A (1974) Lepton number as the fourth "color." Phys Rev D 10:275–289. https://doi.org/10.1103/PhysRevD.10.275

24. Georgi H, Glashow SL (1974) Unity of all elementary-particle forces. Phys Rev Lett 32:438–441. https://doi.org/10.1103/PhysRevLett.32.438

25. Fritzsch H, Minkowski P (1974) Universality of the basic interactions. Phys Lett B 53:373–376. https://doi.org/10.1016/0370-2693(74)90406-7

26. Gross DJ, Wilczek F (1973) Ultraviolet behavior of non-Abelian Gauge theories. Phys Rev Lett 30:1343–1346. https://doi.org/10.1103/PhysRevLett.30.1343

27. Georgi H, Quinn HR, Weinberg S (1974) Hierarchy of interactions in unified gauge theories. Phys Rev Lett 33:451–454. https://doi.org/10.1103/PhysRevLett.33.451

28. Zel'dovich YaB, Khlopov MYu (1978) On the concentration of relic magnetic monopoles in the universe. Phys Lett B 79:239–241. https://doi.org/10.1016/0370-2693(78)90232-0

29. Preskill JP (1979) Cosmological production of superheavy magnetic monopoles. Phys Rev Lett 43:1365–1368. https://doi.org/10.1103/PhysRevLett.43.1365

30. Guth AH (2015) Alan Guth—Session I. In: AIP Oral histories. https://www.aip.org/history-programs/niels-bohr-library/oral-histories/34306-1. Accessed 27 May 2020

31. Guth AH (2015) Alan Guth—Session II. In: AIP Oral histories. https://www. aip.org/history-programs/niels-bohr-library/oral-histories/34306-2. Accessed 27 May 2020
32. Guth AH, Tye S-HH (1980) Phase transitions and magnetic monopole production in the very early universe. Phys Rev Lett 44:631–635. https://doi. org/10.1103/PhysRevLett.44.631
33. Rindler W (1956) Visual horizons in world models. Mon Not R Astron Soc 116:662. https://doi.org/10.1093/mnras/116.6.662
34. Misner CW (1968) The Isotropy of the Universe. Astrophys J 151:431. https://doi.org/10.1086/149448
35. Weinberg S (1972) Gravitation and cosmology: principles and applications of the general theory of relativity, Wiley-VCH, Weinheim, Germany
36. Guth AH (1981) Inflationary universe: a possible solution to the horizon and flatness problems. Phys Rev D 23:347–356. https://doi.org/10.1103/PhysRevD.23.347
37. Starobinskiĭ AA (1979) Spectrum of relict gravitational radiation and the early state of the universe. Sov J Exp Theor Phys Lett 30:682
38. Linde AD (1974) Is the cosmological constant really constant? Pisma v Zhurnal Eksperimentalnoi i Teoreticheskoi Fiziki 19:320–322
39. Coleman S, Weinberg E (1973) Radiative corrections as the origin of spontaneous symmetry breaking. Phys Rev D 7:1888–1910. https://doi.org/10.1103/PhysRevD.7.1888
40. Linde AD (1982) A new inflationary universe scenario: a possible solution of the horizon, flatness, homogeneity, isotropy and primordial monopole problems. Phys Lett B 108:389–393. https://doi.org/10.1016/0370-2693(82)91219-9
41. Albrecht A, Steinhardt PJ (1982) Cosmology for grand unified theories with radiatively induced symmetry breaking. Phys Rev Lett 48:1220–1223. https://doi.org/10.1103/PhysRevLett.48.1220
42. Albrecht A, Steinhardt PJ, Turner MS, Wilczek F (1982) Reheating an inflationary universe. Phys Rev Lett 48:1437–1440. https://doi.org/10.1103/PhysRevLett.48.1437
43. Sakharov AD (1966) The initial stage of an expanding universe and the appearance of a nonuniform distribution of matter. Sov J Exp Theor Phys 22:241
44. Mukhanov VF, Chibisov GV (1981) Quantum fluctuations and a nonsingular universe. Sov J Exp Theor Phys Lett 33:532
45. Mukhanov VF, Chibisov GV (1982) Energy of vacuum and the large-scale structure of the universe. Zhurnal Eksperimentalnoi i Teoreticheskoi Fiziki 83:475–487
46. Starobinsky AA (1982) Dynamics of phase transition in the new inflationary universe scenario and generation of perturbations. Phys Lett B 117:175–178. https://doi.org/10.1016/0370-2693(82)90541-X

47. Guth AH, Pi S-Y (1982) Fluctuations in the new inflationary universe. Phys Rev Lett 49:1110–1113. https://doi.org/10.1103/PhysRevLett.49.1110

48. Hawking SW (1982) The development of irregularities in a single bubble inflationary universe. Phys Lett B 115:295–297. https://doi.org/10.1016/0370-2693(82)90373-2

49. Bardeen JM, Steinhardt PJ, Turner MS (1983) Spontaneous creation of almost scale-free density perturbations in an inflationary universe. Phys Rev D 28:679–693. https://doi.org/10.1103/PhysRevD.28.679

50. Smoot GF, Bennett CL, Kogut A et al (1992) Structure in the COBE differential microwave radiometer first-year maps. Astrophys J Lett 396:L1–L5. https://doi.org/10.1086/186504

51. Guth AH (2007) Eternal inflation and its implications. J Phys A Math Gen 40:6811–6826. https://doi.org/10.1088/1751-8113/40/25/S25

52. Steinhardt PJ (1983) Natural inflation. The very early universe. Cambridge University Press, Cambridge, pp 251–266

53. Vilenkin A (1983) Birth of inflationary universes. Phys Rev D 27:2848–2855. https://doi.org/10.1103/PhysRevD.27.2848

54. Linde AD (1983) Chaotic inflation. Phys Lett B 129:177–181. https://doi.org/10.1016/0370-2693(83)90837-7

55. Linde AD (1986) Eternally existing self-reproducing chaotic inflanationary universe. Phys Lett B 175:395–400. https://doi.org/10.1016/0370-2693(86)90611-8

56. Linde AD (1994) The self-reproducing inflationary universe. Sci Am 271:48–55. https://doi.org/10.1038/scientificamerican1194-48

57. Linde AD (1982) Nonsingular Regenerating Inflationary Universe. Report 82-0554, Cambridge University

58. Ehrenfest P (1918) In that way does it become manifest in the fundamental laws of physics that space has three dimensions? Koninklijke Nederlandse Akademie van Wetenschappen Proc Ser B Phys Sci 20:200–209

59. Dicke RH (1961) Dirac's cosmology and Mach's principle. Nature 192:440–441. https://doi.org/10.1038/192440a0

60. Carter B (1974) Large number coincidences and the anthropic principle in cosmology. In: Confrontation of cosmological theories with observational data. In: Proceedings of the symposium, Krakow, Poland, 10–12 September 1973. D. Reidel Publishing Co, Doredrecht, pp 291–298

61. Carr BJ, Rees MJ (1979) The anthropic principle and the structure of the physical world. Nature 278:605–612. https://doi.org/10.1038/278605a0

62. Rosental IL (1980) On numerical values of fundamental constants. NASA

63. Davies PCW, Unwin SD (1981) Why is the cosmological constant so small. Proc R Soc Lond Ser A 377:147–149. https://doi.org/10.1098/rspa.1981.0119

64. Hawking SW (1982) The cosmological constant and the weak anthropic principle, p 423

65. Weinberg S (1987) Anthropic bound on the cosmological constant. Phys Rev Lett 59:2607–2610. https://doi.org/10.1103/PhysRevLett.59.2607

66. Linde AD (2017) A brief history of the multiverse. Rep Prog Phys 80:022001. https://doi.org/10.1088/1361-6633/aa50e4
67. Riess AG, Filippenko AV, Challis P et al (1998) Observational evidence from supernovae for an accelerating universe and a cosmological constant. Astron J 116:1009–1038. https://doi.org/10.1086/300499
68. Perlmutter S, Aldering G, Goldhaber G et al (1999) Measurements of Ω and Λ from 42 high-redshift supernovae. Astrophys J 517:565–586. https://doi.org/10.1086/307221
69. Martel H, Shapiro PR, Weinberg S (1998) Likely values of the cosmological constant. Astrophys J 492:29–40. https://doi.org/10.1086/305016
70. Polchinski J (2005) String theory. Cambridge University Press, Cambridge, UK
71. Bousso R, Polchinski J (2000) Quantization of four-form fluxes and dynamical neutralization of the cosmological constant. J High Energy Phys 06:006. https://doi.org/10.1088/1126-6708/2000/06/006
72. Douglas MR (2003) The statistics of string/M theory vacua. J High Energy Phys 05:046. https://doi.org/10.1088/1126-6708/2003/05/046
73. Susskind L (2003) The anthropic landscape of string theory, p 26
74. Kachru S, Kallosh R, Linde A et al (2003) Towards inflation in string theory. J Cosmol Astropart Phys 10:013. https://doi.org/10.1088/1475-7516/2003/10/013
75. Susskind L (2005) The cosmic landscape: string theory and the illusion of intelligent design, 1st edn. Little, Brown and Company, New York
76. Gross DJ (2005) Where do we stand in fundamental (string) theory. Physica Scripta Volume T 117:102–105. https://doi.org/10.1238/Physica.Topical.117a00102
77. Gross D (2005) The future of physics. Int J Mod Phys A 20:5897–5909. https://doi.org/10.1142/S0217751X05029095
78. Steinhardt PJ (2011) The inflation debate. Sci Am 304:36–43. https://doi.org/10.1038/scientificamerican0411-36
79. Feeney SM, Johnson MC, Mortlock DJ, Peiris HV (2011) First observational tests of eternal inflation: analysis methods and WMAP 7-year results. Phys Rev D 84:043507. https://doi.org/10.1103/PhysRevD.84.043507
80. Aguirre A, Johnson MC (2011) A status report on the observability of cosmic bubble collisions. Rep Prog Phys 74:074901. https://doi.org/10.1088/0034-4885/74/7/074901
81. Kleban M (2011) Cosmic bubble collisions. Class Quant Grav 28:204008. https://doi.org/10.1088/0264-9381/28/20/204008
82. Dirac PAM (1937) The cosmological constants. Nature 139:323. https://doi.org/10.1038/139323a0
83. Dirac PAM (1938) A new basis for cosmology. Proc R Soc Lond Ser A 165:199–208. https://doi.org/10.1098/rspa.1938.0053

84. Moffat JW (1993) Superluminary universe: a possible solution to the initial value problem in cosmology. Int J Mod Phys D 2:351–365. https://doi.org/10.1142/S0218271893000246
85. Moffat JW (1993) Quantum gravity, the origin of time and time's arrow. Found Phys 23:411–437. https://doi.org/10.1007/BF01883721
86. Albrecht A, Magueijo J (1999) Time varying speed of light as a solution to cosmological puzzles. Phys Rev D 59:043516. https://doi.org/10.1103/PhysRevD.59.043516
87. Magueijo J (2003) New varying speed of light theories. Rep Prog Phys 66:2025–2068. https://doi.org/10.1088/0034-4885/66/11/R04
88. Magueijo J, Smolin L (2004) Gravity's rainbow. Class Quant Grav 21:1725–1736. https://doi.org/10.1088/0264-9381/21/7/001
89. Awad A, Farag Ali A, Majumder B (2013) Nonsingular rainbow universes. J Cosmol Astropart Phys 10:052. https://doi.org/10.1088/1475-7516/2013/10/052
90. Alexander S, Barrow JD, Magueijo J (2016) Turning on gravity with the Higgs mechanism. Class Quant Grav 33:14LT01. https://doi.org/10.1088/0264-9381/33/14/14LT01
91. Magueijo J (2009) Bimetric varying speed of light theories and primordial fluctuations. Phys Rev D 79:043525. https://doi.org/10.1103/PhysRevD.79.043525
92. Afshordi N, Magueijo J (2016) Critical geometry of a thermal big bang. Phys Rev D 94:101301. https://doi.org/10.1103/PhysRevD.94.101301
93. Ashtekar A (1986) New variables for classical and quantum gravity. Phys Rev Lett 57:2244–2247. https://doi.org/10.1103/PhysRevLett.57.2244
94. Rovelli C, Smolin L (1988) Knot theory and quantum gravity. Phys Rev Lett 61:1155–1158. https://doi.org/10.1103/PhysRevLett.61.1155
95. Rovelli C, Smolin L (1990) Loop space representation of quantum general relativity. Nucl Phys B 331:80–152. https://doi.org/10.1016/0550-3213(90)90019-A
96. Ashtekar A, Rovelli C, Smolin L (1992) Weaving a classical metric with quantum threads. Phys Rev Lett 69:237–240. https://doi.org/10.1103/PhysRevLett.69.237
97. Rovelli C, Smolin L (1995) Spin networks and quantum gravity. Phys Rev D 52:5743–5759. https://doi.org/10.1103/PhysRevD.52.5743
98. Rovelli C, Smolin L (1995) Discreteness of area and volume in quantum gravity. Nucl Phys B 442:593–619. https://doi.org/10.1016/0550-3213(95)00150-Q
99. Bojowald M (2000) Loop quantum cosmology: I Kinematics. Class Quant Grav 17:1489–1508. https://doi.org/10.1088/0264-9381/17/6/312
100. Bojowald M (2000) Loop quantum cosmology: II. Volume operators. Class Quant Grav 17:1509–1526. https://doi.org/10.1088/0264-9381/17/6/313

101. Bojowald M (2001) Loop quantum cosmology: III. Wheeler-DeWitt operators. Class Quant Grav 18:1055–1069. https://doi.org/10.1088/0264-9381/18/6/307

102. Bojowald M (2001) Loop quantum cosmology: IV. Discrete time evolution. Class Quant Grav 18:1071–1087. https://doi.org/10.1088/0264-9381/18/6/308

103. Bojowald M (2001) Absence of a singularity in Loop Quantum cosmology. Phys Rev Lett 86:5227–5230. https://doi.org/10.1103/PhysRevLett.86.5227

104. Bojowald M (2008) Loop Quantum Cosmology. Living Reviews in Relativity 11:4. https://doi.org/10.12942/lrr-2008-4

105. Ashtekar A, Singh P (2011) Loop quantum cosmology: a status report. Class Quant Grav 28:213001. https://doi.org/10.1088/0264-9381/28/21/213001

106. Bojowald M (2007) What happened before the Big Bang? Nat Phys 3:523–525. https://doi.org/10.1038/nphys654

107. Ashtekar A, Gupt B (2017) Quantum gravity in the sky: interplay between fundamental theory and observations. Class Quant Grav 34:014002. https://doi.org/10.1088/1361-6382/34/1/014002

108. Khoury J, Ovrut BA, Steinhardt PJ, Turok N (2001) Ekpyrotic universe: colliding branes and the origin of the hot big bang. Phys Rev D 64:123522. https://doi.org/10.1103/PhysRevD.64.123522

109. Steinhardt PJ, Turok N (2005) The cyclic model simplified. NewAR 49:43–57. https://doi.org/10.1016/j.newar.2005.01.003

110. Linde AD (2003) Inflationary theory versus the ekpyrotic/cyclic scenario. Fut Theor Phys Cosmol 801–838

111. Linde AD (2014) Inflationary cosmology after Planck 2013. arXiv e-prints arXiv:1402.0526

112. Wilczek F (2016) Unification of force and substance. Philos Trans Royal Soc A: Math Phys Eng Sci 374:20,150,257. https://doi.org/10.1098/rsta.2015.0257, CC BY 4.0

113. Franknoi A, Morrison D, Wolff SC (2016) Astronomy, Houston, Texas. Access for free at https://openstax.org/books/astronomy/pages/1-introduction, CC BY 4.0)

114. Mughal MZ, Ahmad I, García Guirao JL (2021) Relativistic Cosmology with an Introduction to Inflation. Universe 7:276. https://doi.org/10.3390/univer se7080276 , CC BY 4.0)

5

The Dark Universe

In Chap. 4, we looked at how the cosmic microwave background draws a veil across the early universe, and how scientists have tried to peer past that barrier by extrapolating known physics back as far as it will go. We also saw how they ended up with a radically new theory for what the Big Bang was—the end of inflation—and for what cosmic origins might really entail: Namely, an eternal and incomprehensibly infinitely multiverse.

In this chapter, we'll follow what theorists and observers have learned on this side of the veil—starting with their discovery of just how much "known physics" leaves out. Indeed, by pioneering ever more precise and sophisticated versions of their standard tools—spectroscopy, the cosmic distance ladder, and observations of the cosmic microwave background (CMB)—these researchers have uncovered at least two new phenomena that have shaped the evolution of the universe in profound ways, yet are utterly mysterious and absolutely invisible.

5.1 Dark Matter

In the late 1960s and early 1970s, as researchers inspired by the discoveries of quasars and the CMB came flocking into the newly fashionable field of cosmology, they focused much of their attention on the ultimate fate of the universe. Will it expand forever? Or will it eventually halt its expansion and crash back down to a Big Bang in reverse—the Big Crunch?

M. M. Waldrop, *Cosmic Origins*, https://doi.org/10.1007/978-3-030-98214-0_5

According to the cosmological equations derived from general relativity by Friedmann and Lemâitre, the answer was intimately tied to one number: the average density of matter in the universe. If the average was at or below a certain critical density, equivalent to a few hydrogen atoms per cubic meter, the collective gravitational pull of all those galaxies would not be enough to reverse the expansion, and the universe would indeed keep growing forever. (It would also be infinite in extent.) But if the average density was greater than that critical value, the universe would be spherical and finite—albeit very big. And it would one day face the Big Crunch.

Unfortunately, the average cosmic density was a huge question mark. Most cosmologists were rooting for a high number on purely philosophical grounds: the notion of a closed, bounded universe seemed a lot more reassuring than the prospect of a cosmos stretching out to infinity [1]. Yet the available evidence said otherwise. As nearly as anyone could tell from counting galaxies and estimating their masses, the average density of the universe was just a fraction of the critical value, which would imply a universe that was both infinite and eternal [2–6].

This disconnect was a puzzle, to put it mildly. As the particle physicist Steven Weinberg put it in his 1972 textbook on cosmology, if you believed in a closed universe, "the mass density ... must be found somewhere outside the normal galaxies. But where?" [5] And with that question now surging to the fore, cosmologists began to look back with fresh eyes on observational anomalies that had been piling up for decades [1, 7].

5.1.1 Missing Mass

The Cluster Conundrum

The first of these anomalies had been uncovered all the way back in 1933, by Swiss-American astronomer Fritz Zwicky [8]. His starting point was the landmark survey of galactic redshifts that Edwin Hubble and Milton Humason had published just two years earlier (see Chap. 2). Zwicky had been particularly struck by the fact that several galaxies in their sample lay in a dense grouping known as the Coma Cluster (Fig. 5.1), and showed quite a lot of scatter around the straight-line velocity-distance relation [9]. To Zwicky, this "noise" was the signal: Since the Coma galaxies were essentially the same distance from Earth, he reasoned, the redshift scatter must arise from their *peculiar* velocities—that is, their random motions within the cluster. Furthermore, Zwicky realized, those peculiar velocities must be kept in check by the galaxies' mutual gravitational attraction; otherwise, the Coma galaxies would

Fig. 5.1 The Coma Cluster, where Fritz Zwicky discovered the first hints of cosmic dark matter. (Credit: NASA, ESA, J. Mack (STScI), and J. Madrid (Australian Telescope National Facility))

have flown apart, and the cluster would have evaporated eons ago. Since it obviously had not, Zwicky reasoned, he could take those peculiar velocities to work backwards and estimate the cluster's total mass.

And therein lay the anomaly: When Zwicky estimated the Coma Cluster's mass based on the 1,000 km/s peculiar velocities in Hubble and Humason's data, he got one number. But when he did another estimate based on what astronomers could actually see—roughly 800 galaxies in the cluster, times about a billion solar masses per galaxy—he got a smaller number. *Much* smaller: if that visible mass was all there was, Zwicky calculated, then the random speed of a typical Coma galaxy wouldn't be much more than about 80 *km/s*—less than one-tenth the actual value. So to him, the conclusion was inescapable: the Coma galaxies were being held together by an enormous amount of gravity coming from something astronomers *couldn't* see—something that Zwicky, writing in German, called *dunkle Materie*: Dark matter.

Zwicky wasn't actually the first scientist to use this phrase; the honor probably goes to the French mathematician and physicist Henri Poincaré, who had already discussed the possibility of a universe filled with *matière obscure* in 1906 [10]. But neither scientist was using "dark matter" in its modern sense. Poincaré had assumed that it consisted of dim or burnt-out stars, or dense clouds of interstellar gas and dust—ordinary stuff that just happened to be dark. So, it seems, did Zwicky: He said as much in 1937, when he published a much more detailed analysis of the Coma Cluster that again found anomalously high peculiar velocities [11]. And so did astronomer Sinclair Smith, who in 1936 had found similar speeds in another dense grouping, the Virgo Cluster [12]. And in truth, it was a natural assumption: Dim stars and gas clouds were at least known to exist, even if their actual abundance was unknown.

But still—1,000 km/s? Velocities more than 10 times higher than expected?

Most astronomers found these cluster numbers to be so big and so baffling that they just figured the analysis had gone wrong somewhere—even if the alternatives didn't make much sense either. For example, some tried to argue that the Coma Cluster actually *was* evaporating, appearances to the contrary, so there was no need to assume that gravity was holding the galaxies back, and no need for dark matter. But that argument fell apart as soon as you asked why we're still seeing the Coma cluster today, billions of years after the universe began. If it's not gravitationally bound, why didn't it disintegrate eons ago? [7]

Flat Rotation Curves?

Meanwhile, other astronomers had been looking for dark matter in individual spiral galaxies. Although they didn't often communicate with their colleagues working on clusters, their strategy was essentially the same: use the velocities you *can* see to measure the mass that you (mostly) *can't*. In the case of spirals, which were known to be rotating like pinwheels (see Chap. 2), this meant using spectroscopic red- and blueshifts to get the rotation speed of stars lying further and further out in the spiral arms. Then Newton's law of gravity would relate the speed at any given radius to the distribution of mass in the galaxy.[1]

In practice this was tricky, since the outer parts of spiral arms tended to be so faint it was hard to get good spectra. And even when astronomers did manage to get them, their results were … strange. The expectation had always been that the rotational velocities would fall with distance like they do in the solar system, where virtually all the mass is concentrated in the Sun, and where close-in planets like Earth orbit much faster than outer planets like Jupiter and Saturn. Since spiral galaxies likewise seemed to concentrate most of their mass at the center, in bright, fuzzy "bulges" that contained most of

---

[1] Of course, a completely up-to-date analysis would require Einstein's theory of gravity, general relativity. But Newton's venerable inverse-square law turns out to be an excellent approximation, since the stars in any normal galaxy move at velocities far less than the speed of light.

the starlight, the physics ought to be the same. But the actual observations didn't seem to give a consistent answer. A few observers even claimed to find rotation curves that didn't fall at all. So again, most astronomers in the field were content to wait for better technology and better data.

They got both in a rush, starting in 1970. In a study published in February of that year, American astronomers Vera Rubin and Kent Ford described how they had used a new, ultra-sensitive spectrograph developed by Ford to obtain a high-resolution rotation curve for the Andromeda galaxy. Their data encompassed essentially the entire visible disk, and showed that the rotational velocities out in the spiral arms weren't falling as fast with radius as might be expected. In fact, just as the arms were tailing off into invisibility, there were signs that the curve might be flattening out [13].

Or maybe not. The observations were still iffy, so Rubin and Ford were careful not to draw any direct conclusions about the distribution of dark stuff in Andromeda [1]. But a few months later the Australian astronomer Ken Freeman did take that leap, in the appendix of a paper devoted to the statistical analysis of star abundance in the spiral arms [14]. Instead of Andromeda, Freeman focused on two nearby galaxies known as M33 and NGC 300. And instead of using spectroscopy at visible wavelengths, he got his velocities from already-published radio observations—specifically, from shifts in a strong spectral line emitted by interstellar hydrogen gas at a wavelength of 21 cm. Since this 21-cm line allowed him to see motions in the gas even when it wasn't being lit by stars, he could construct rotation curves extending way beyond the visible disk. The spatial resolution wasn't great, Freeman admitted, but the rotation curves definitely kept rising even as the distribution of stars was tailing off—which implied that the total mass density was *not* tailing off. "*If [these data] are correct,*" he wrote, "*then there must be in these galaxies additional matter which is undetected, either optically or at 21 cm.*"

Given the many uncertainties, though, Freeman was almost as tentative in this assertion as Rubin and Ford had been. And so, for a time, were the radio astronomers who jumped in to settle this question with their own observations [15–18]. Viewed in retrospect, the 21-cm studies they produced in the early 1970s were making a stronger and stronger case for spiral-galaxy rotation curves that became flat with increasing radius—meaning, once you worked backwards from Newton's Law, that each spiral galaxy was embedded in a massive, halo of this invisible stuff that extended far beyond the visible stars. Yet the authors continued to stress the uncertainties, if only because it was still far from clear in the early 1970s that a massive halo was the *only* explanation.

The Case for Dark Matter

That began to change in 1974, when two independent groups finally put the anomalies seen in the spiral-galaxy rotation curves together with those seen in the clusters, and suggested that both might have the same explanation: Dark matter. The Princeton team, a collaboration between physicist James Peebles and astronomers Jeremiah Ostriker and Amos Yahil, memorably made the case in their opening sentence: "*There are reasons, increasing in number and quality, to believe that the masses of ordinary galaxies may have been underestimated by a factor of 10 or more.*" Furthermore, both they and the second team, led by Estonian astronomer Jaan Einasto, calculated that this unknown stuff comprised at least 20% of the mass needed to close the universe—and conceivably might comprise all of it [19, 20].

There were still plenty of skeptics even then. But most cosmologists embraced the idea: Their missing mass had at last been found. And the observational evidence for large amounts of dark matter continued to pour in [21, 22]. The stuff turned up in every galaxy that astronomers examined, whether in visible light or at the 21-cm radio frequencies.

There was theoretical support, as well—notably from computer simulations that Peebles and Ostriker had carried out a year before their paper with Yahil, and that had helped inspire it. Their simulations showed that spiral-galaxy pinwheels held together by gravity were highly unstable, and would quickly crumple into a bar-like or elliptical structure—unless they happened to be embedded in a much larger, invisible halo of dark matter whose gravity kept the instability in check. Since spiral galaxies were everywhere, and seemed to be billions of years old, the obvious implication was that every one of them had a dark-matter halo to stabilize it and ensure its survival. And if that were true, then dark matter had to be just as ubiquitous as the spirals were [23].

So by 1979, when American astronomers Sandra Faber and Jay Gallagher published a lucid and influential review [24] of the evidence for dark matter, their conclusion was already gaining widespread acceptance: "*the case for invisible mass in the universe is very strong and becoming stronger.*"

But of course, that conclusion just led to another obvious question: If dark matter is ubiquitous yet utterly invisible—what *is* it?

5.1.2   MACHOs, MOND, or WIMPs?

The answer wasn't quite so obvious. Four decades earlier, Zwicky, Smith and others had simply assumed that the dark stuff they were detecting was some non-luminous form of ordinary matter much like the stars and nebulae they had seen before. And even with flat rotation curves and all the rest, many astronomers in the 1970s would have been happy to keep making that assumption [7]. But as the 1970s turned into the 1980s, this notion was becoming harder and harder to sustain.

It isn't Gas

Interstellar gas wouldn't work, for example. At the scale of clusters, observers looking at radio, optical, and x-ray wavelengths had indeed been able to find a faint haze of ionized hydrogen drifting in the space between galaxies—so much gas, in fact, that it added up to a lot more mass than in the cluster galaxies themselves. But this was still not nearly enough mass to explain Zwicky's anomaly [25–27]. And likewise with individual galaxies. Any spiral galaxy will have lots of gas clouds lining its arms, where the dense knots of interstellar hydrogen and helium are a breeding ground for new stars. These clouds are often quite dark—in visible light. But they light up like Broadway when they are viewed at the 21-cm radio wavelength of neutral hydrogen. And again, surveys at that wavelength showed that there were not nearly enough hydrogen clouds in the spiral arms to account for dark matter. And in any case, the velocity anomalies were also turning up in big elliptical galaxies that had very little gas and no arms at all.

MACHOs?

Nor could dark matter be in the form of massive astrophysical compact halo objects, or MACHOs—a grab-bag category of objects that shared little except for being dense, dark, and infinitesimally tiny on a galactic scale. Astronomers knew of many candidates, both real and hypothetical. The list included dim red stars; old, burnt-out white-dwarf stars; faintly glowing brown dwarf stars without quite enough mass to ignite thermonuclear fusion; neutron stars formed from the cores of detonating supernovae; free-roaming, Jupiter-sized planets—even primordial black holes left over from the Big Bang. And although it was hard to see how any known astrophysical process could have produced enough of these compact objects[28], they would remain viable dark-matter candidates well into the 1990s and 2000s.

MACHOs would finally fall from favor only after three separate collaborations spent those years trying and failing to find them via "microlensing," a general-relativistic phenomenon in which the gravitational field of a MACHO drifting between us and a distant star will briefly focus the star's light and make it seem brighter [29–31]. The researchers did detect a few such objects—after monitoring tens of millions of stars for the better part of a decade. But again, not nearly enough [32–34].

And finally, if the lack of observational evidence wasn't sufficient, there was a compelling cosmological argument that a group of French and American physicists had first put forward in 1973 [35]. If you take the observed cosmic abundances of hydrogen, deuterium, and helium, they noted, then you can work backwards through all the nuclear reactions that produced these isotopes during the Big Bang, and get an estimate of how much ordinary, baryonic matter the universe started with. (*Baryonic*, from a Greek word meaning heavy, was a term that physicists had started to use for any type of matter containing protons, neutrons, or more complex atomic nuclei—a category that includes gas clouds, stars, planets, and us.) When the researchers did this, they found that the density of baryonic matter couldn't be any more than 10% of the cosmic critical density. This was less than half the *lower* limit for dark matter that the Princeton and Estonian teams would find from cluster velocities and rotation curves just a year later.

In short, it was clear even then that there had to be an immense amount of dark matter that wasn't baryonic. And as time went on, the constraints would only get stronger; the best figures today show that the average cosmic density of baryonic matter is just under 1/5 the average density of dark matter [36].

But if not baryonic matter, then what?

MOND?

Maybe nothing. In 1963, the Italian physicist Arrigo Finzi broached this possibility in a remarkably prescient paper that turned the cluster-velocity question on its head: Instead of using the anomalous motions plus Newton's inverse-square law of gravity as evidence for a lot of stuff nobody could see, Finzi used the apparent lack of stuff to argue for a different law of gravity [37]. After all, he pointed out, if gravity were stronger at large distances than Newton's formula predicted, then you could explain the anomalous motions without any dark matter whatsoever. (Finzi also anticipated many of the arguments against baryonic dark matter that would only become widespread later.)

Finzi's suggestion was generally ignored at the time, perhaps because it was too weird for physicists in 1963. But two decades later, the Israeli physicist Mordehai Milgrom got considerably more attention when he introduced a formula for Modified Newtonian Dynamics, or MOND, and used it to give a dark-matter-free account of flat rotation curves and the like [38–40]. Admittedly, Milgrom's first version of the idea was crude and ad-hoc. But he and the Mexican–American physicist Jacob Bekenstein would spend the next two decades developing it into a much more sophisticated theory—a complicated variant of Einstein's general relativity [41–43].

The MOND approach continued to give a reasonably good account of anomalous motions in individual galaxies. Where it struggled, however, was with clusters of galaxies—especially when it came to the increasingly detailed data obtained via the gravitational lensing effect, which is a macro version of the microlensing discussed above.

The idea that gravity can act as a lens dated back to Einstein himself, who pointed out in 1936 that it followed from his theory of general relativity. Starting with his insight that massive bodies like the Sun will bend the surrounding space–time and deflect any passing starlight—the prediction that was so spectacularly confirmed during the solar eclipse of 1919—Einstein concluded that the deflection would end up magnifying and brightening any star, nebula or galaxy lying in the background [29]. True, the resulting images would be distorted, fragmented, or smeared into arcs centered on the foreground object; as lenses go, gravity is terrible. And even with the most massive stars, Einstein calculated, the effect would be infinitesimal. But then, as he cheerfully admitted in his paper, he was sharing the lensing idea simply as an amusing thought experiment, not as something that he thought astronomers could hope to see.

Zwicky, however, was not so sure. Einstein had been thinking of lensing by individual stars in his 1936 paper. But Zwicky quickly realized that, if big galaxy clusters like Coma and Virgo were really as massive as the peculiar velocities implied, then the lensing effect would be far bigger—and in principle, observable. Better still, measurements of how the images of far-distant galaxies are displaced, brightened, and distorted would allow astronomers to work backwards in any given cluster, and get an estimate of its mass that was completely independent of the velocity data. [11, 44, 45]. If the two mass estimates agreed, then dark matter was probably real. If they didn't—well, those velocities would need another explanation.

In practice, this test of the dark-matter idea was difficult to pull off, if only because those distant galaxies are so faint. The first clear example of a gravitational lens wouldn't be detected until 1979, when astronomers found a pair of neighboring quasars that had exactly the same redshifts and spectra— a coincidence that was ludicrously implausible unless a galaxy lying along the line of sight just happened to be creating two different images of the same object [46]. And it wasn't until the late 1980s that improved instrumentation revealed that the mysterious blue arcs seen in many clusters were in fact what Zwicky had predicted: the lensed images of distant galaxies [47–50].

After that, though, the evidence accumulated rapidly—especially once the Hubble Space Telescope was able to gather ultra-clear imagery of the clusters, some of which turned out to contain blue arcs by the dozens. Ground-based observations also improved dramatically in the 1990s, thanks to new "adaptive optics" technology that minimized distortions produced by the atmosphere. By 2000, lensing had been detected in most large clusters, and astronomers were working backwards to reconstruct the distribution of mass with increasing confidence [51–54] (Fig. 5.2).

The upshot was that Zwicky had been right: clusters contain a *lot* more mass than can be seen in the galaxies, in roughly a five-to-one ratio of dark matter to baryonic matter. And more than that, dark matter isn't centered on individual galaxies. It fills up the entire cluster, with its densest knots often showing up in empty stretches where there's no galaxy to be seen.

This last finding was an uncomfortable one for the MOND approach, since even a modified gravitational field would presumably be centered on the visible mass. And in 2004, the MOND theory's prospects went from bad to worse when astronomers got their first close look at the Bullet Cluster: a remote system of galaxies that had been discovered only in 1998, and that proved to be *two* clusters just emerging from a violent, 100-million-year long collision [55] (Fig. 5.3). Optical images from Hubble and from ground-based telescopes showed that the visible galaxies had suffered minimal damage. Galaxies are so small from a cosmic perspective that head-on encounters are rare, so the two groups had mostly flown right through one another and started to separate again. At the same time, however, the intergalactic gas clouds that carry most of the two clusters' baryons had hit hard. Tenuous or not, the hydrogen ions had collided often enough to exert a drag force on the clouds, slowing them down and creating dramatic shock waves. High-resolution X-ray images captured by NASA's Chandra satellite showed that the two gas clouds were also starting to separate again, but were trailing well behind the visible galaxies.

The real eye-opener, however, was the dark-matter map of the Bullet Cluster created from gravitational lensing data. It showed two lobes of dark stuff aligned along the same axis as the gas clouds, but centered much further out—right on top of the two visible clusters, in fact. The dark-matter clumps, along with the visible galaxies, had clearly passed right though one another feeling no friction at all—even as the clouds of baryonic gas were suffering a trainwreck. The obvious conclusion was that dark matter and baryonic matter are two separate things that can move independently—a fact that was virtually impossible to square with any form of modified gravity. [56, 57]. And in the years since then, as astronomers keep finding the same separation between

Fig. 5.2 The invisible dark matter that permeates cluster Abell 370 acts as a gravitational lens, magnifying and distorting the images of more distant galaxies lying behind the cluster. Among these distorted images is the dramatic yellow arc just below the cluster's center, as well as a multitude of much fainter blue arcs. By measuring these lensed images, cosmologists can work backwards and map where the cluster's dark matter actually is; their results are plotted here as a blue haze. (Credit: NASA, ESA, D. Harvey (École Polytechnique Fédérale de Lausanne, Switzerland), R. Massey (Durham University, UK), the Hubble SM4 ERO Team and ST-ECF. CC BY-SA 3.0 IGO)

dark matter and baryonic matter in every colliding cluster they look at, that conclusion has only grown stronger [58].

Or WIMPs?

Even as the Bullet Cluster discovery was all but ruling out the modified-gravity interpretation of dark matter, however, it was reinforcing what many

Fig. 5.3 The Bullet Cluster shows a clear separation between the baryonic matter seen via x-rays (pink) and the dark matter mapped via gravitational lensing (blue). (Credit: X-ray: NASA/CXC/M.Markevitch et al.; Optical: NASA/STScI; Magellan/U.Arizona/D.Clowe et al.; Lensing Map: NASA/STScI; ESO WFI; Magellan/U.Arizona/D.Clowe et al. CC BY-SA 3.0 IGO)

cosmologists had long since embraced as the most plausible explanation—that dark matter is actually a swarm of elementary particles left over from the Big Bang [59].

If nothing else, this notion made dark matter's exceedingly strange behavior seem sensible. For example, the stuff's complete and utter invisibility could be explained quite naturally if the particles were electrically neutral, and thus allowed photons to pass through without getting scattered or absorbed. Likewise with dark matter's ability to flow through stars, planets, people, and other blobs of dark matter without even slowing down: This simply meant that the particles didn't respond to the strong force, either. (Responding to the weak force was optional.) And finally, since the Big Bang presumably produced *lots* of these particles, dark matter's immense gravitational pull would follow if they had even a tiny mass.

Particles existed with each of these properties. And one family of them, the neutrinos, had them all: zero charge, tiny masses, no strong interactions, the works. Also, since neutrinos are not protons or neutrons, they are most definitely not baryonic—which is why, for a brief time in the 1970s and early

1980s, physicists could hope that they had already found the dark-matter particle [60–64].

It was not to be. By the mid-1980s, for reasons explained below, researchers had been forced to conclude that all-neutrino dark matter was inconsistent with the observed distribution of galaxies and clusters in the universe. But neutrinos' brief reign did encourage astronomers and physicists to keep looking for other non-baryonic dark-matter particles that could solve the dark matter conundrum.

Fortunately, there were lots of other candidate particles—hypothetical, to be sure, but well-motivated by physicists' continued work on the unified field theories discussed in Chap. 4. One such particle, the axion, was proposed in the late 1970s as a fix for the theory of quarks, gluons, and the strong interactions—a.k.a. quantum chromodynamics, or QCD. Without the axion, the equations of QCD contained a loophole that allowed the strong force to treat particles and antiparticles differently—a behavior known as *CP violation* that experiments had ruled out to a very high precision. With the axion, that loophole vanished and QCD was no longer out of step with the data [65–68].

Although the axion's mass and interaction strength were not well-defined by QCD, astronomers were soon able to put stringent limits on both. If axions were too massive, to take just one example, they would cause the exploding stars known as supernovae to cool off faster than the measured rate. Taken together, observations like these showed that axions—if they existed— would have to be very light (10^{-6} to 10^{-4} electron Volts, or less than a billionth the mass of an electron), and very feeble in their interactions with other particles. But they would also be stable and copiously produced during the Big Bang—meaning that they might well account for the dark matter we see today [7].

The axion has thus been a favorite for this role ever since—but not the only favorite. During those same years, for example, physicists had begun to explore theories in which our space–time had more than the usual four dimensions, but with the extras curled up so tight that they were imperceptible. The effect would be to give every known particle a series of higher-mass partners, in somewhat the same way that an organ pipe produces a series of higher-pitched harmonics. When researchers worked through the details, several of these partners looked like good dark matter candidates [69].

Meanwhile, another large family of dark matter candidates had turned up as physicists developed supersymmetry—a mathematically elegant, but decidedly non-intuitive approach that calls for enlarging Einstein's space–time with new dimensions that behave something like the square root of an ordinary dimension. [70–72]. Fortunately, supersymmetry's practical effect was quite straightforward: Each of our familiar particles would get paired with a super-partner that had the same electric charge, but an internal angular momentum, or "spin," that differed by one-half unit. So spin-½ particles like electrons, quarks, and neutrinos would be partnered with spin-0 selectrons, squarks, and sneutrinos, respectively. Likewise, spin-1 force particles such as photons and gluons would be partnered with spin-½ photinos and gluinos; the spin-0 Higgs with a spin-½ Higgsino; and even the spin-2 graviton (a quantum of gravity) with a spin-3/2 gravitino.

By the early 1980s, physicists had developed supersymmetric versions of both general relativity and the standard model [73–75]. By the mid-1980s they were looking at the gravitino, the photino, and all the other electri-cally neutral superpartners as serious dark matter candidates [76–79]. And by decade's end, when all these candidates had received the collective nick-name WIMPs—weakly interacting, massive particles [80]—most researchers had embraced them as the most likely answer to the dark-matter puzzle.

The only trick was to prove it.

5.1.3   The Large-Scale Structure of the Universe

Efforts to accomplish that feat were already proceeding on two broad fronts. One, discussed in the next chapter, included the many different (and so far, unsuccessful) attempts to detect dark matter particles directly, either in the lab or in the sky.

The other, less direct but more fruitful, was to test the WIMP theory with computer modeling. Could it really account for the universe we see around us?

This kind of modeling certainly wasn't a new ambition; attempts to simu-late galaxies and other gravitating systems dated back to the 1940s. But the sheer scale of the dark-matter problem made the simulations daunting. To do it right, you'd have to build a computer model that incorporated every-thing known about baryonic matter, general relativity, cosmic expansion and all the rest, and then add in your hypothetical WIMPS. Next, you'd have to let the machine calculate step by step how gravity tugged on the more-or-less uniform haze of dark and ordinary matter emerging from Big Bang, and slowly turned them into a clumpy network of galaxies and clusters. And finally, after billions of simulated years had ticked down to the simulated present day, you'd have compare its predicted distribution of galaxies and dark matter halos to the real universe.

In any previous decade, cosmological modeling on this scale would have been out of the question. Fortunately, however, four major developments converged in the 1980s to make the simulations dramatically more powerful [7].

Advances in Simulation

The first and most obvious advance was the rapid growth in computer power, coupled with researchers' increasingly sophisticated modeling techniques. In the 1960s they would have been lucky to simulate the motion of 100 mass points (roughly analogous to stars.) By the 1980s they could handle millions—still just a fraction of the billions of mass points used in today's simulations, but enough to yield useful results.

Second was the community's embrace of the cosmic inflation theory (see Chap. 4). Inflation gave modelers a specific prediction for the overall shape of the universe—namely, that it had to be geometrically flat to a very high precision—as well as for its matter content: The total density of baryonic matter plus dark matter had to be *exactly* equal to the critical density.[2] Inflation theory also gave modelers a crucial piece of input: their models of galaxy formation should start with the inflation-era quantum fluctuations first calculated by Linde, Guth, and others in the early 1980s. These size and location of these fluctuations were random, but their statistical distribution was quite predictable. So the question became whether gravity could start with fluctuations distributed in this way, and turn them into today's distribution of galaxies.

A third development was astronomers' growing appreciation of dark matter itself. Whatever this stuff was, it was so massive that galaxy formation had little to do with the visible galaxies. The comparatively thin wisps of baryonic gas—mostly hydrogen and helium—couldn't do much more than follow along with whatever the dark matter was doing. Presumably this gas would eventually collect in the densest clumps of dark matter and form galaxies there. And that was good news for the modelers, since they could predict where the galaxies and clusters would be without having to simulate friction, shock waves, heating, or any of the other complexities that baryonic matter is heir to, but dark matter isn't.

And finally, modelers could now test their output against dramatic new data on how galaxies and clusters were actually organized in the universe.

---

[2] Or at least, that was the assumption at the time. As we'll see later in this chapter, however, there was another invisible component to consider.

The Cosmic Web

Astronomers had known for decades that this "large-scale structure" was at least somewhat hierarchical: Stars formed galaxies; galaxies formed groups (a prime example being our own local group, which includes the Milky Way, Andromeda, and several others); and groups formed clusters like Coma or Virgo. But there were thought to be few if any connections among the largest clusters, which seemed like isolated islands poking up from the Pacific.

The first hint of something more came in 1982, when a team at the Harvard-Smithsonian Center for Astrophysics published the first large, 3D map of the nearby universe constructed from redshift data on 2,200 galaxies [81]. The authors described the galaxy distribution they found as "frothy," as if they were looking at a cross-sectional slice of soap bubbles—except that here, the bubble walls were made of clusters and superclusters, and the bubble interiors were galaxy-free voids more than a hundred million light-years across. Subsequent 3D maps created by groups such as the Center for Astrophysics, the 2-Degree Field consortium headquartered in Australia, and the Sloan Digital Sky Survey would push out to much greater distances, and would harvest redshifts from many more galaxies. But they would show essentially the same frothy structure—often referred to as the cosmic web (Fig. 5.4).

Even in 1982, however, it was clear even to the authors of that first survey that the existing simulations of galaxy formation were missing something crucial: When they compared their data to one of the best models then available [82], the differences were glaring. That model had clusters—but there were no filaments connecting the clusters, and no cosmic web.

The missing ingredient, obviously, was dark matter: Those early simulations had left it out. But there was a crucial caveat. Within a year, a team of physicists from the University of California at Berkeley had modeled the formation of the cosmic web that *did* include dark matter—but a specific type of dark matter consisting of massive neutrinos, which was then the most popular WIMP candidate. The neutrinos just didn't work [83, 84]. The Berkeley team got dark matter clumping at the largest cosmic scales. But down at the level of galaxies and clusters, they just got a washed-out continuum. Neutrino dark matter, they concluded, "appears to be ruled out."

In retrospect, the source of the problem was obvious: Neutrinos emerging from the Big Bang would still be moving at virtually the speed of light, making them far too "hot" to condense into small clumps. Thus the washed-out effect. Much better were axions, photinos, or any other dark-matter particle that would now be "cold"—that is, moving at non-relativistic veloc-ities. For cosmological purposes, it almost didn't matter what these cold

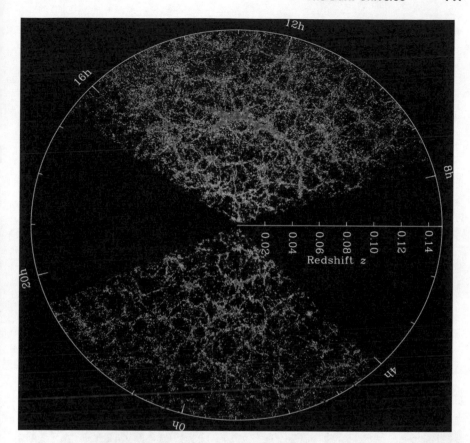

Fig. 5.4 A modern map of the cosmic web, created by the Sloan Digital Sky Survey. Each dot is a galaxy. The two dark wedges are sections of the distant universe that are obscured by interstellar gas and dust in our own galaxy. (Credit: M. Blanton and SDSS, CC-BY)

particles were. Starting in 1982, several groups had published pencil-and-paper calculations suggesting that cold dark matter would naturally form clumps of just the right size to be galaxies (or globular star clusters), and that these clumps would naturally coalesce over time into clusters, filaments, and all the rest [85–89]. Then in 1984, those calculations were confirmed when modelers put the math into a computer and completed the first cosmic simulation using cold dark matter: They found that such non-relativistic particles would produce a large-scale web very much like the one being revealed in the redshift surveys [90] (Fig. 5.5).

All of which is why the cold dark matter (CDM) paradigm has reigned ever since as *the* dominant paradigm for explaining the formation of galaxies, clusters, and the large-scale structure of the universe [7].

Albeit, with the addition of one *other* missing piece...

Fig. 5.5 Modern simulations of the cosmic web of galaxies, like this one, look very much like the real web—*if* they incorporate cold dark matter. (Credit: CLUES—Constrained Local Universe Evolution Simulation, CC BY 4.0)

5.2   Dark Energy

Although most cosmologists felt there was clearly something *right* about the CDM model, at least some of them had suspected from the beginning that it couldn't be the whole story.

If you took the model at face value, for example, you'd just set the baryonic fraction of the universe to its observed value—roughly 5% of the critical density—and then let cold dark matter alone make up the other 95%. Indeed, many of the early models did just that [91]. But 95% CDM was a stretch, since most observations pointed to a total density of baryonic matter plus CDM closer to 20%. Also, the pure CDM simulations tended to produce more clustering on small scales than the observations showed, and less on large scales.

Modelers could and did try to address these issues by fiddling with the parameters of the theory. But that just left them with other problems, such as a model universe only about 10 billion years old—younger than some stars [92]. So by the mid-1980s, many theorists and modelers were asking how to get that CDM fraction down without undermining its success in the simulations.

One obvious solution was to mix in some hot dark matter—that is, neutrinos. Neutrinos had the virtue of actually existing. And simulations based on this Hot–Cold Dark Matter model did fit the galaxy surveys pretty well [91, 93]. For better or worse, however, this option eventually fell out of favor—not least because it became apparent that standard-model neutrinos aren't massive enough to come anywhere close to that 20% figure [94].

Another, much more promising approach was to say that the baryonic-plus-CDM fraction was just what it seemed to be—roughly 20%—and to forget about making up the difference. Embrace the possibility that cosmic density might be way below critical. Of course, going this route meant accepting the idea of an open, infinite universe that's not even close to flat, and giving up on the whole inflation thing. But simulations based on this Open-CDM assumption worked pretty well [95]. So as one review put it, keep an open mind! [96].

5.2.1 Was Einstein Right?

Finally, though, there was a possibility that seemed so unappealing that people hesitated to even suggest it: What if Einstein had been right all those years ago, and the universe really *did* have a non-zero cosmological constant? [97, 98].

The unappealing part wasn't the mathematical argument—although that was certainly counterintuitive. Naively, for example, you'd think that the cosmological constant was the opposite of mass: It would produce a kind of relentless, outward pressure pushing every point in the universe away from its neighbors. That's why Einstein had invented the constant in the first place, to balance the inward force of gravity and keep his model universe from collapsing (see Chap. 2). But when you dug into things a little deeper, it turned out that a nonzero cosmological constant effectively gave every point in the universe a certain amount of energy—which, by $E = mc^2$, was equivalent to saying that there would be a certain density of mass in even the remotest regions of empty space. Because this mass density wasn't attached to any actual particle, moreover—it would be a *constant*, fixed equally at every point and unable to move—it would contribute a big chunk of the critical density without having the slightest effect on galaxy clustering. And because the critical density was equivalent to only a few atoms per cubic meter, on the average, the constant could achieve this feat while still being small enough to have escaped detection.

So yes, the equations worked. But that was the *un*appealing part. This approach patched up the problems with one mystery—cold dark matter—by invoking another mystery: the constant. The whole thing was messy and ugly.

The particle physicists were even more skeptical. In their world, the cosmological constant was a major irritant. It cropped up everywhere in their theories, in the form of quantum field fluctuations that threatened to fill empty space with energies an estimated 10^{120} times larger than anything the cosmologists were talking about. Physicists could be happy if the constant exactly equaled zero, since they could achieve that simply by invoking supersymmetry: The vacuum energy produced by the field fluctuations of each particle would be exactly canceled by vacuum fluctuations of its superpartner. Indeed, this had always been major argument for believing in supersymmetry. But trying to explain a cancellation that was *almost* perfect, yet missed by one part in 10^{120}—well, that was *really* ugly [99, 100].

So cosmology entered the 1990s with Open-CDM model looking like the model to beat[92]. But that's also when the balance of opinion began to shift. In a landmark study published in 1990, a team of British astronomers scanned 20 years' worth of astronomical imagery from Australia, and compiled a map of some 2 million galaxies spanning much of the southern hemisphere [101]. What they found was that the real universe shows a lot more clustering at the largest scales than any pure CDM simulation could account for—and that only one fix was truly consistent with the data: "a positive cosmological constant could solve many of the problems of the standard CDM model and should be taken seriously."[102].

This forthright, data-backed declaration led many other cosmologists (although by no means all) to start doing just that [91, 92, 103, 104]. Since Einstein and everyone who came after him had denoted the cosmological constant by the Greek letter lambda (λ, capital form Λ), this scenario soon became known as the lambda cold dark matter model: ΛCDM. And modelers soon found that a constant equivalent to maybe 60 to 70% of the critical density, with a dark-plus-baryonic matter fraction making up the rest, would bring the simulated cosmic web almost perfectly in line with the observations [105–110].

This line of thinking was reinforced even further in 1995, when Ostriker and Steinhardt published a short, but influential review that brought together data on the Hubble parameter, the age of the universe and every other observational test of the models. When you put it all together, the authors found, everything pointed in the same direction—toward a universe best described by the ΛCDM model, not Open CDM [92, 111].

And then came 1998.

5.2.2 A Shift in the Paradigm

Supernovae

Even as the modelers had been arguing about CDM variants, it turned out, two rival teams of astronomers had been racing to perfect their observations of type 1a supernovae: a class of exploding stars that promised to become a whole new kind of standard candle. Like the Cepheid variables that had played this role for nearly a century, supernovas of this type were reasonably common in the universe (a few per galaxy per century), and offered an easy way for astronomers to measure a given explosion's intrinsic luminosity—which in this case turned out to be strongly correlated with its "light curve," a plot of how its brightness waxes and wanes over the course of days and weeks [112].

Unlike the Cepheids, however, type 1a supernovae were bright enough to be seen across billions of light years. And in 1998, after meticulously refining their calibrations of the supernovae and then plotting the redshift-distance relation for dozens of them, both teams announced the same jaw-dropping result: the Hubble expansion of the universe was accelerating. And more than that, the observed acceleration was just about what you'd expect from a cosmological constant corresponding to roughly 70% of the critical density, and a dark-plus-baryonic matter fraction around 30% [112–114] (Fig. 5.6).

It was this dramatic revelation that really pushed cosmologists toward a broader acceptance of the ΛCDM idea—although they weren't always happy about it. The model was still asking them to embrace two mysteries instead of one. It was still forcing particle physicists to deal with that 10^{120} business. And it left everyone wondering what Λ actually *was*. An honest-to-Einstein constant that affected every point in the universe equally? Some new kind of quantum field—a "quintessence" that could vary (very slowly) from place to place and moment to moment [115]? Something even weirder [116]? No one knew—then or now. The best anyone could do was cover all the possibilities with a catchy new name for Λ: "dark energy" [117].

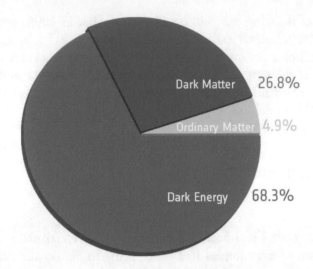

Fig. 5.6 In 2013, data from the European Space Agency's Planck satellite yielded this updated estimate for the composition of the universe. The ordinary matter that comprises stars, galaxies, planets, and us is dwarfed by dark energy and dark matter. (Credit: ESA and the Planck Collaboration, CC BY-SA 3.0 IGO)

Yet there it was: the dark energy/ΛCDM scenario was the only one left that fit *all* the data. And for any remaining doubters, this scenario soon got an even more spectacular confirmation from a completely different source.

Cosmic Sound Waves

The idea went back at least as far as 1970, when Soviet and American physicists independently realized that the hot plasma of the early universe must have been humming with sound waves—literally [118, 119].

Why? The key insight was that baryons in the plasma would constantly be trying to form clumps because of gravity, while the surrounding firestorm of photons would constantly be trying to push them apart again. The resulting compression-pushback cycle would have been very much like what happens with sound waves in Earth's atmosphere, just on a far larger scale. So the cosmic plasma would have been crisscrossed with very similar waves, which came to be known as *baryon acoustic oscillations*.

Extending this analogy with ordinary sound even further, later calculations suggested that the expanding cosmos would have acted on these oscillations something like an organ pipe, enhancing those wavelengths that resonated while damping those that didn't [120]. True, the oscillations themselves would have disappeared during recombination, since that's when the plasma turned into a bunch of neutral atoms and the photons flew away to become the CMB. But the organ-pipe effect would have lived on in those CMB temperature fluctuations—the same ones that presumably provided the seeds for galaxy formation (see Chap. 4). So a statistical analysis on the angular size of these CMB fluctuations ought to reveal a series of peaks corresponding to the resonant wavelengths [121]. And those peaks, in turn, should provide a wealth of information about the cosmic conditions that formed them, including the Hubble parameter, the cosmic matter content, and whether or not the universe is flat.

If, of course, you could get the data. This baryon acoustic oscillation approach was hypothetical until NASA launched its Cosmic Background Explorer satellite (COBE) in 1989. A year later, COBE returned the first really precise measurement of the CMB temperature, which turned out to be 2.735 ± 0.06K [122]. And in 1992, COBE finally detected those elusive temperature variations—or *anisotropies*, as they are known in the trade [123]. The anisotropies turned out to be tiny, only a few parts in 100,000. And COBE's microwave vision was much too fuzzy to see any of the baryon acoustic peaks. In the *multipole* units l that most observers preferred—a system in which larger numbers denote smaller features—COBE couldn't do anything smaller than about $l = 25$, or roughly 7°. That's 14 times the width of the full moon; the peaks were expected to be no more than a quarter that size, or even smaller. That would put them somewhere beyond $l = 100$, equivalent to a degree or two.

But COBE did show that the anisotropies existed. By decade's end, finer-grained microwave data obtained from balloons, sounding rockets, and ground-based observatories in the Andes had begun to show hints of at a peak around $l = 200$, or about 1°—pretty much where the ΛCDM model predicted it would be [124]. In 2000, two long-duration balloon experiments, BOOMERanG in Antarctica and MAXIMA in Texas, convincingly measured that peak and beyond [125–128].

Then in 2001, NASA launched the Wilkinson Microwave Anisotropy Probe, or WMAP.[3] WMAP scanned the entire microwave sky with 33 times the resolution of COBE and 45 times its sensitivity [129]. During its nine years of operation, it would produce a series of spectacular portraits of the CMB anisotropies, each more refined and detailed than the last—along with an equally spectacular series of statistical plots showing not just the first

---

[3] Originally, NASA just called the spacecraft MAP. The agency renamed it in 2003 to honor Princeton physicist David Wilkinson, who—as we saw in Chap. 3—had been one of the first to confirm the existence of the cosmic microwave background back in the 1960s. Wilkinson had gone on to be one of the prime movers behind both COBE and MAP before his 2002 death from cancer.

Fig. 5.7 WMAP's best view of temperature fluctuations in the cosmic microwave background. The color-coding represents deviations from the CMB's 2.725 K average, and ranges from black on the cool end, through blue, green, and yellow, to red on the warm end. The deviations themselves are tiny, amounting to no more than a few millionths of a degree. Yet they correspond to primordial variations in the matter density that would later be magnified by gravity, and would eventually become the galaxies and clusters we see today. (Credit: NASA / WMAP Science Team)

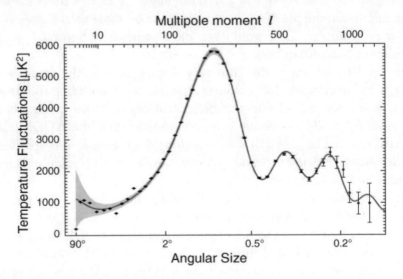

Fig. 5.8 WMAP's best view of the CMB's baryon acoustic peaks. The red line is the best-fit prediction of the ΛCDM model. (Credit: NASA / WMAP Science Team)

acoustic peak but at least two higher harmonics [130–136]. The peaks fit beautifully with a ΛCDM universe (Figs. 5.7 and 5.8).

Down on the ground, meanwhile, astronomers were extracting an alternative view of the baryon acoustic oscillations from the large-scale galaxy surveys mentioned above—notably the Sloan Digital Sky Survey. The researchers' thinking was that the cosmic web of galaxies and clusters would still show signs of the peaks and valleys that the oscillations had imposed on the CMB-era density anisotropies, even after billions of years of gravitational collapse. And they were right: When SDSS and its fellow surveys did a statistical analysis of the cosmic web, they consistently found that the resonance peaks and other cosmological parameters fit very well with the ΛCDM predictions [137–142].

The Sloan survey is still ongoing—the project launched Phase V in 2020—but WMAP is not. The spacecraft was turned off in 2010 following the previous year's launch of its successor, the European Space Agency's Planck spacecraft. Planck is a more ambitious mission that boasted a more extensive suite of instruments, three times better angular resolution and—thanks to detectors cooled to within a degree of absolute zero—a much higher sensitivity to subtle temperature variations. During the four-plus years it operated before its coolant ran out, Planck produced its own series of spectacular results, including a multipole plot showing many acoustic peaks. But the fundamental conclusion was unchanged: aside from one nagging discrepancy discussed in the next chapter, the ΛCDM model describes the universe beautifully; none of the other models come anywhere close.

References

1. de Swart JG, Bertone G, van Dongen J (2017) How dark matter came to matter. Nat Astron 1:0059. https://doi.org/10.1038/s41550-017-0059
2. Noonan TW (1971) The mean cosmic density from galaxy counts and mass data. Publ Astron Soc Pac 83:31. https://doi.org/10.1086/129059
3. Shapiro SL (1971) The density of matter in the form of galaxies. Astron J 76:291. https://doi.org/10.1086/111122
4. Peebles PJE (1971) Physical cosmology. Princeton University Press, Princeton, N.J.
5. Weinberg S (1972) Gravitation and cosmology: principles and applications of the general theory of relativity
6. Gott JR III, Gunn JE, Schramm DN, Tinsley BM (1974) An unbound universe. Astrophys J 194:543–553. https://doi.org/10.1086/153273

7. 7.Bertone G, Hooper D (2018) History of dark matter. Rev Mod Phys 90:045002.https://doi.org/10.1103/RevModPhys.90.045002
8. Zwicky F (1933) Die Rotverschiebung von extragalaktischen Nebeln. Helvetica Phys Acta 6:110–127
9. Hubble E, Humason ML (1931) The velocity-distance relation among extra-galactic nebulae. Astrophys J 74:43. https://doi.org/10.1086/143323
10. Poincare H (1906) The milky way and the theory of gases. Pop Astron 14:475–488
11. Zwicky F (1937) On the masses of nebulae and of clusters of nebulae. Astrophys J 86:217. https://doi.org/10.1086/143864
12. Smith S (1936) The mass of the Virgo cluster. Astrophys J 83:23. https://doi.org/10.1086/143697
13. Rubin VC, Ford WK Jr (1970) Rotation of the Andromeda nebula from a spectroscopic survey of emission regions. Astrophys J 159:379. https://doi.org/10.1086/150317
14. Freeman KC (1970) On the disks of spiral and S0 galaxies. Astrophys J 160:811. https://doi.org/10.1086/150474
15. Rogstad DH, Shostak GS (1972) Gross properties of five Scd galaxies as deter-mined from 21-CENTIMETER observations. Astrophys J 176:315. https://doi.org/10.1086/151636
16. Whitehurst RN, Roberts MS (1972) High-velocity neutral hydrogen in the central region of the andromeda galaxy. Astrophys J 175:347. https://doi.org/10.1086/151562
17. Roberts MS, Rots AH (1973) Comparison of rotation curves of different galaxy types. Astron Astrophys 26:483–485
18. Roberts MS (1975) The Rotation Curve of Galaxies. 69:331
19. Ostriker JP, Peebles PJE, Yahil A (1974) The size and mass of galaxies, and the mass of the universe. Astrophys J Lett 193:L1–L4. https://doi.org/10.1086/181617
20. Einasto J, Kaasik A, Saar E (1974) Dynamic evidence on massive coronas of galaxies. Nature 250:309–310. https://doi.org/10.1038/250309a0
21. Rubin VC, Ford WK Jr, Thonnard N (1978) Extended rotation curves of high-luminosity spiral galaxies. IV—systematic dynamical properties. SA through SC. Astrophys J Lett 225:L107–L111. https://doi.org/10.1086/182804
22. Bosma A (1978) The distribution and kinematics of neutral hydrogen in spiral galaxies of various morphological types
23. Ostriker JP, Peebles PJE (1973) A numerical study of the stability of flattened galaxies: or, can cold galaxies survive? Astrophys J 186:467–480. https://doi.org/10.1086/152513
24. Faber SM, Gallagher JS (1979) Masses and mass-to-light ratios of galaxies. Annu Rev Astron Astrophys 17:135–187. https://doi.org/10.1146/annurev.aa.17.090179.001031

25. Penzias AA (1961) Free hydrogen in the Pegasus I cluster of galaxies. Astron J 66:293. https://doi.org/10.1086/108555

26. Woolf NJ (1967) On the stabilization of clusters of galaxies by ionized gas. Astrophys J 148:287. https://doi.org/10.1086/149148

27. Meekins JF, Fritz G, Chubb TA, Friedman H (1971) Physical sciences: X-rays from the coma cluster of galaxies. Nature 231:107–108. https://doi.org/10.1038/231107a0

28. Hegyi DJ, Olive KA (1983) Can galactic halos be made of baryons? Phys Lett B 126:28–32. https://doi.org/10.1016/0370-2693(83)90009-6

29. Einstein A (1936) Lens-like action of a star by the deviation of light in the gravitational field. Science 84:506–507. https://doi.org/10.1126/science.84.2188.506

30. Irwin MJ, Webster RL, Hewett PC et al (1989) Photometric variations in the Q2237 + 0305 system—first detection of a microlensing event. Astron J 98:1989–1994. https://doi.org/10.1086/115272

31. Gould A (2000) A natural formalism for microlensing. Astrophys J 542:785–788. https://doi.org/10.1086/317037

32. Alcock C, Allsman RA, Alves DR et al (2000) The MACHO project: microlensing results from 5.7 years of large Magellanic cloud observations. Astrophys J 542:281–307. https://doi.org/10.1086/309512

33. Lasserre T, Afonso C, Albert JN et al (2000) Not enough stellar mass Machos in the Galactic halo. Astron Astrophys 355:L39–L42

34. Tisserand P, Le Guillou L, Afonso C et al (2007) Limits on the macho content of the Galactic Halo from the EROS-2 survey of the Magellanic clouds. Astron Astrophys 469:387–404. https://doi.org/10.1051/0004-6361:20066017

35. Reeves H, Audouze J, Fowler WA, Schramm DN (1973) On the origin of light elements. Astrophys J 179:909–930. https://doi.org/10.1086/151928

36. Planck Collaboration, Aghanim N, Akrami Y et al (2018) Planck 2018 results. VI. Cosmological parameters. ArXiv E-Prints 1807:arXiv:1807.06209

37. Finzi A (1963) On the validity of Newton's law at a long distance. Mon Not R Astron Soc 127:21. https://doi.org/10.1093/mnras/127.1.21

38. Milgrom M (1983) A modification of the Newtonian dynamics as a possible alternative to the hidden mass hypothesis. Astrophys J 270:365–370. https://doi.org/10.1086/161130

39. Milgrom M (1983) A modification of the Newtonian dynamics—implications for galaxies. Astrophys J 270:371–383. https://doi.org/10.1086/161131

40. Milgrom M (1983) A Modification of the newtonian dynamics—implications for galaxy systems. Astrophys J 270:384. https://doi.org/10.1086/161132

41. Bekenstein J, Milgrom M (1984) Does the missing mass problem signal the breakdown of Newtonian gravity? Astrophys J 286:7–14. https://doi.org/10.1086/162570

42. Bekenstein JD (1988) Phase coupling gravitation: symmetries and gauge fields. Phys Lett B 202:497–500. https://doi.org/10.1016/0370-2693(88)91851-5

43. Bekenstein JD (2004) Relativistic gravitation theory for the modified Newtonian dynamics paradigm. Phys Rev D 70:083509.https://doi.org/10.1103/PhysRevD.70.083509

44. Zwicky F (1937) Nebulae as gravitational lenses. Phys Rev 51:290–290. https://doi.org/10.1103/PhysRev.51.290

45. Zwicky F (1937) On the probability of detecting nebulae which act as gravitational lenses. Phys Rev 51:679–679. https://doi.org/10.1103/PhysRev.51.679

46. Walsh D, Carswell RF, Weymann RJ (1979) 0957 + 561 A, B—Twin quasistellar objects or gravitational lens. Nature 279:381–384. https://doi.org/10.1038/279381a0

47. Lynds R, Petrosian V (1986) Giant luminous arcs in galaxy clusters. 18:1014

48. Soucail G, Fort B, Mellier Y, Picat JP (1987) A blue ring-like structure, in the center of the A 370 cluster of galaxies. Astron Astrophys 172:L14–L16

49. Kovner I (1987) Giant luminous arcs from gravitational lensing. Nature 327:193–194. https://doi.org/10.1038/327193c0

50. Lynds R, Petrosian V (1989) Luminous arcs in clusters of galaxies. Astrophys J 336:1–8. https://doi.org/10.1086/166989

51. Tyson JA, Valdes F, Wenk RA (1990) Detection of systematic gravitational lens galaxy image alignments—mapping dark matter in galaxy clusters. Astrophys J Lett 349:L1–L4. https://doi.org/10.1086/185636

52. Kaiser N, Squires G (1993) Mapping the dark matter with weak gravitational lensing. Astrophys J 404:441–450. https://doi.org/10.1086/172297

53. Mellier Y (1999) Probing the universe with weak lensing. Annu Rev Astron Astrophys 37:127–189. https://doi.org/10.1146/annurev.astro.37.1.127

54. Massey R, Kitching T, Richard J (2010) The dark matter of gravitational lensing. Rep Prog Phys 73:086901.https://doi.org/10.1088/0034-4885/73/8/086901

55. Tucker W, Blanco P, Rappoport S, et al (1998) 1E 0657–56: a contender for the hottest known cluster of galaxies. Astrophys J Lett 496:L5–L8. https://doi.org/10.1086/311234

56. Markevitch M, Gonzalez AH, Clowe D et al (2004) Direct constraints on the dark matter self-interaction cross section from the merging galaxy cluster 1E 0657–56. Astrophys J 606:819–824. https://doi.org/10.1086/383178

57. Clowe D, Bradač M, Gonzalez AH et al (2006) A direct empirical proof of the existence of dark matter. Astrophys J Lett 648:L109–L113. https://doi.org/10.1086/508162

58. Harvey D, Massey R, Kitching T et al (2015) The nongravitational interactions of dark matter in colliding galaxy clusters. Science 347:1462–1465. https://doi.org/10.1126/science.1261381

59. Feng JL (2010) Dark matter candidates from particle physics and methods of detection. Annu Rev Astron Astrophys 48:495–545. https://doi.org/10.1146/annurev-astro-082708-101659

60. Gershtein SS, Zel'dovich YaB (1966) Rest mass of Muonic neutrino and cosmology. Sov J Exp Theor Phys Lett 4:120–122

61. Szalay AS, Marx G (1976) Neutrino rest mass from cosmology. Astron Astrophys 49:437–441

62. Lee BW, Weinberg S (1977) Cosmological lower bound on heavy-neutrino masses. Phys Rev Lett 39:165–168. https://doi.org/10.1103/PhysRevLett.39.165

63. Gunn JE, Lee BW, Lerche I et al (1978) Some astrophysical consequences of the existence of a heavy stable neutral lepton. Astrophys J 223:1015–1031. https://doi.org/10.1086/156335

64. Doroshkevich AG, Zel'dovich YaB, Syunyaev RA, Khlopov MY (1980) Astrophysical implications of the neutrino rest mass. II–The density-perturbation spectrum and small-scale fluctuations in the microwave background. III—Nonlinear growth of perturbations and the missing mass. Pisma V Astron Zhurnal 6:457–469

65. Peccei RD, Quinn HR (1977) CP conservation in the presence of pseudoparticles. Phys Rev Lett 38:1440–1443. https://doi.org/10.1103/PhysRevLett.38.1440

66. Peccei RD, Quinn HR (1977) Constraints imposed by CP conservation in the presence of pseudoparticles. Phys Rev D 16:1791–1797. https://doi.org/10.1103/PhysRevD.16.1791

67. Wilczek F (1978) Problem of strong P and T invariance in the presence of instantons. Phys Rev Lett 40:279–282. https://doi.org/10.1103/PhysRevLett.40.279

68. Weinberg S (1978) A new light boson? Phys Rev Lett 40:223–226. https://doi.org/10.1103/PhysRevLett.40.223

69. Bertone G, Hooper D, Silk J (2005) Particle dark matter: evidence, candidates and constraints. Phys Rep 405:279–390. https://doi.org/10.1016/j.physrep.2004.08.031

70. Gervais J-L, Sakita B (1971) Field theory interpretation of supergauges in dual models. Nucl Phys B 34:632–639. https://doi.org/10.1016/0550-3213(71)90351-8

71. Gol'Fand YuA, Likhtman EP (1971) Extension of the algebra of Poincare group generators and violation of p invariance. Sov J Exp Theor Phys Lett 13:323

72. Wess J, Zumino B (1974) Supergauge transformations in four dimensions. Nucl Phys B 70:39–50. https://doi.org/10.1016/0550-3213(74)90355-1

73. Freedman DZ, van Nieuwenhuizen P, Ferrara S (1976) Progress toward a theory of supergravity. Phys Rev D 13:3214–3218. https://doi.org/10.1103/PhysRevD.13.3214

74. Deser S, Zumino B (1976) Consistent supergravity. Phys Lett B 62:335–337. https://doi.org/10.1016/0370-2693(76)90089-7

75. Dimopoulos S, Georgi H (1981) Softly broken supersymmetry and SU(5). Nucl Phys B 193:150–162. https://doi.org/10.1016/0550-3213(81)90522-8

76. Pagels H, Primack JR (1982) Supersymmetry, cosmology, and new physics at teraelectronvolt energies. Phys Rev Lett 48:223–226. https://doi.org/10.1103/PhysRevLett.48.223

77. Weinberg S (1983) Upper bound on gauge-fermion masses. Phys Rev Lett 50:387–389. https://doi.org/10.1103/PhysRevLett.50.387

78. Goldberg H (1983) Constraint on the photino mass from cosmology. Phys Rev Lett 50:1419–1422. https://doi.org/10.1103/PhysRevLett.50.1419

79. Ellis J, Hagelin JS, Nanopoulos DV et al (1984) Supersymmetric relics from the big bang. Nucl Phys B 238:453–476. https://doi.org/10.1016/0550-3213(84)90461-9

80. Steigman G, Turner MS (1985) Cosmological constraints on the properties of weakly interacting massive particles. Nucl Phys B 253:375–386. https://doi.org/10.1016/0550-3213(85)90537-1

81. Davis M, Huchra J, Latham DW, Tonry J (1982) A survey of galaxy redshifts. II—the large scale space distribution. Astrophys J 253:423–445. https://doi.org/10.1086/159646

82. Efstathiou G, Eastwood JW (1981) On the clustering of particles in an expanding universe. Mon Not R Astron Soc 194:503–525. https://doi.org/10.1093/mnras/194.3.503

83. Frenk CS, White SDM, Davis M (1983) Nonlinear evolution of large-scale structure in the universe. Astrophys J 271:417–430. https://doi.org/10.1086/161209

84. White SDM, Frenk CS, Davis M (1983) Clustering in a neutrino-dominated universe. Astrophys J Lett 274:L1–L5. https://doi.org/10.1086/184139

85. Peebles PJE (1982) Large-scale background temperature and mass fluctuations due to scale-invariant primeval perturbations. Astrophys J Lett 263:L1–L5. https://doi.org/10.1086/183911

86. Bond JR, Szalay AS, Turner MS (1982) Formation of galaxies in a gravitino-dominated universe. Phys Rev Lett 48:1636–1639. https://doi.org/10.1103/PhysRevLett.48.1636

87. Blumenthal GR, Pagels H, Primack JR (1982) Galaxy formation by dissipationless particles heavier than neutrinos. Nature 299:37. https://doi.org/10.1038/299037a0

88. Peebles PJE (1984) Dark matter and the origin of galaxies and globular star clusters. Astrophys J 277:470–477. https://doi.org/10.1086/161714

89. Blumenthal GR, Faber SM, Primack JR, Rees MJ (1984) Formation of galaxies and large-scale structure with cold dark matter. Nature 311:517–525. https://doi.org/10.1038/311517a0

90. Davis M, Efstathiou G, Frenk CS, White SDM (1985) The evolution of large-scale structure in a universe dominated by cold dark matter. Astrophys J 292:371–394. https://doi.org/10.1086/163168

91. Bertschinger E (1998) Simulations of structure formation in the universe. Annu Rev Astron Astrophys 36:599–654. https://doi.org/10.1146/annurev.astro.36.1.599

92. Calder L, Lahav O (2010) Dark energy: how the paradigm shifted. Phys World 23:32–37. https://doi.org/10.1088/2058-7058/23/01/33
93. Liddle AR, Lyth DH, Schaefer RK et al (1996) Pursuing parameters for critical-density dark matter models. Mon Not R Astron Soc 281:531. https://doi.org/10.1093/mnras/281.2.531
94. Spergel D (1997) Particle dark matter, 221–240
95. Cole S, Weinberg DH, Frenk CS, Ratra B (1997) Large-scale structure in COBE-normalized cold dark matter cosmogonies. Mon Not R Astron Soc 289:37–51. https://doi.org/10.1093/mnras/289.1.37
96. Coles P, Ellis G (1994) The case for an open Universe. Nature 370:609–615. https://doi.org/10.1038/370609a0
97. Peebles PJE (1984) Tests of cosmological models constrained by inflation. Astrophys J 284:439–444. https://doi.org/10.1086/162425
98. Turner MS, Steigman G, Krauss LM (1984) Flatness of the universe—reconciling theoretical prejudices with observational data. Phys Rev Lett 52:2090–2093. https://doi.org/10.1103/PhysRevLett.52.2090
99. Carroll SM, Press WH, Turner EL (1992) The cosmological constant. Annu Rev Astron Astrophys 30:499–542. https://doi.org/10.1146/annurev.aa.30.090192.002435
100. Carroll S (2002) Dark energy and the preposterous universe, 2
101. Maddox SJ, Sutherland WJ, Efstathiou G, Loveday J (1990) The APM galaxy survey. I—APM measurements and star-galaxy separation. Mon Not R Astron Soc 243:692–712
102. Efstathiou G, Sutherland WJ, Maddox SJ (1990) The cosmological constant and cold dark matter. Nature 348:705–707. https://doi.org/10.1038/348705a0
103. Davis M, Efstathiou G, Frenk CS, White SDM (1992) The end of cold dark matter? Nature 356:489–494. https://doi.org/10.1038/356489a0
104. Ostriker JP (1993) Astronomical tests of the cold dark matter scenario. Annu Rev Astron Astrophys 31:689–716. https://doi.org/10.1146/annurev.aa.31.090193.003353
105. Martel H (1991) N-body simulation of large-scale structures in Lambda not = 0 Friedmann models. Astrophys J 366:353–383. https://doi.org/10.1086/169570
106. Suginohara T, Suto Y (1992) Properties of galactic halos in spatially flat universes dominated by cold dark matter—effects of nonvanishing cosmological constant. Astrophys J 396:395–410. https://doi.org/10.1086/171727
107. Cen R, Gnedin NY, Ostriker JP (1993) A hydrodynamic treatment of the cold dark matter cosmological scenario with a cosmological constant. Astrophys J 417:387. https://doi.org/10.1086/173320
108. Cen R, Ostriker JP (1994) X-ray clusters in a cold dark matter + lambda universe: a direct, large-scale, high-resolution, hydrodynamic simulation. Astrophys J 429:4–21. https://doi.org/10.1086/174297

109. Gnedin NY (1996) Galaxy Formation in a CDM + Lambda Universe. I. Properties of gas and galaxies. Astrophys J 456:1. https://doi.org/10.1086/176623

110. Gnedin NY (1996) Galaxy Formation in a CDM + Lambda Universe. II. Spatial distribution of gas and galaxies. Astrophys J 456:34. https://doi.org/10.1086/176624

111. Ostriker JP, Steinhardt PJ (1995) The observational case for a low-density Universe with a non-zero cosmological constant. Nature 377:600–602. https://doi.org/10.1038/377600a0

112. Kirshner RP (1999) Supernovae, an accelerating universe and the cosmological constant. Proc Natl Acad Sci 96:4224–4227. https://doi.org/10.1073/pnas.96.8.4224

113. Riess AG, Filippenko AV, Challis P et al (1998) Observational evidence from supernovae for an accelerating universe and a cosmological constant. Astron J 116:1009–1038. https://doi.org/10.1086/300499

114. Perlmutter S, Aldering G, Goldhaber G et al (1999) Measurements of Ω and Λ from 42 high-redshift supernovae. Astrophys J 517:565–586. https://doi.org/10.1086/307221

115. Steinhardt PJ, Caldwell RR (1998) Introduction to quintessence. 151:13

116. Caldwell R (2004) Dark energy. In: Phys World. https://physicsworld.com/a/dark-energy/. Accessed 14 Jul 2020

117. Turner MS (1999) Dark matter and dark energy in the Universe. 165:431

118. Sunyaev RA, Zel'dovich YaB (1970) Small-Scale fluctuations of relic radiation. Astrophys Space Sci 7:3–19.https://doi.org/10.1007/BF00653471

119. Peebles PJE, Yu JT (1970) Primeval adiabatic perturbation in an expanding universe. Astrophys J 162:815. https://doi.org/10.1086/150713

120. Hu W, Sugiyama N, Silk J (1997) The physics of microwave background anisotropies. Nature 386:37–43. https://doi.org/10.1038/386037a0

121. Tegmark M (1997) How to measure CMB power spectra without losing information. Phys Rev D 55:5895–5907. https://doi.org/10.1103/PhysRevD.55.5895

122. Mather JC, Cheng ES, Eplee RE Jr et al (1990) A preliminary measurement of the cosmic microwave background spectrum by the Cosmic Background Explorer (COBE) satellite. Astrophys J Lett 354:L37–L40. https://doi.org/10.1086/185717

123. Smoot GF, Bennett CL, Kogut A et al (1992) Structure in the COBE differential microwave radiometer first-year maps. Astrophys J Lett 396:L1–L5. https://doi.org/10.1086/186504

124. Miller AD, Caldwell R, Devlin MJ et al (1999) A Measurement of the angular power spectrum of the cosmic microwave background from L = 100 to 400. Astrophys J Lett 524:L1–L4. https://doi.org/10.1086/312293

125. de Bernardis P, Ade PAR, Bock JJ et al (2000) A flat Universe from high-resolution maps of the cosmic microwave background radiation. Nature 404:955–959. https://doi.org/10.1038/35010035

126. Hanany S, Ade P, Balbi A et al (2000) MAXIMA-1: A measurement of the cosmic microwave background anisotropy on angular scales of 10'–5°. Astrophys J Lett 545:L5–L9. https://doi.org/10.1086/317322

127. Balbi A, Ade P, Bock J et al (2000) Constraints on cosmological parameters from MAXIMA-1. Astrophys J Lett 545:L1–L4. https://doi.org/10.1086/317323

128. MacTavish CJ, Ade PAR, Bock JJ et al (2006) Cosmological parameters from the 2003 flight of BOOMERANG. Astrophys J 647:799–812. https://doi.org/10.1086/505558

129. Bennett CL, Bay M, Halpern M et al (2003) The microwave anisotropy probe mission. Astrophys J 583:1–23. https://doi.org/10.1086/345346

130. Spergel DN, Verde L, Peiris HV et al (2003) First-Year Wilkinson microwave anisotropy probe (WMAP) observations: determination of cosmological parameters. Astrophys J Suppl Ser 148:175–194. https://doi.org/10.1086/377226

131. Spergel DN, Bean R, Doré O et al (2007) Three-Year Wilkinson microwave anisotropy probe (WMAP) observations: implications for cosmology. Astrophys J Suppl Ser 170:377–408. https://doi.org/10.1086/513700

132. Hinshaw G, Weiland JL, Hill RS et al (2009) Five-Year Wilkinson microwave anisotropy probe observations: data processing, sky maps, and basic results. Astrophys J Suppl Ser 180:225–245. https://doi.org/10.1088/0067-0049/180/2/225

133. Bennett CL, Hill RS, Hinshaw G et al (2011) Seven-year Wilkinson microwave anisotropy probe (WMAP) observations: are there cosmic microwave background anomalies? Astrophys J Suppl Ser 192:17. https://doi.org/10.1088/0067-0049/192/2/17

134. Komatsu E, Smith KM, Dunkley J et al (2011) Seven-year Wilkinson microwave anisotropy probe (WMAP) observations: cosmological interpretation. Astrophys J Suppl Ser 192:18. https://doi.org/10.1088/0067-0049/192/2/18

135. Larson D, Dunkley J, Hinshaw G et al (2011) Seven-year Wilkinson microwave anisotropy probe (WMAP) observations: power spectra and WMAP-derived parameters. Astrophys J Suppl Ser 192:16. https://doi.org/10.1088/0067-0049/192/2/16

136. Bennett CL, Larson D, Weiland JL et al (2013) Nine-year Wilkinson microwave anisotropy probe (WMAP) observations: final maps and results. Astrophys J Suppl Ser 208:20. https://doi.org/10.1088/0067-0049/208/2/20

137. Percival WJ, Baugh CM, Bland-Hawthorn J et al (2001) The 2dF galaxy redshift survey: the power spectrum and the matter content of the universe. Mon Not R Astron Soc 327:1297–1306. https://doi.org/10.1046/j.1365-8711.2001.04827.x

138. Percival WJ, Sutherland W, Peacock JA et al (2002) Parameter constraints for flat cosmologies from cosmic microwave background and 2dFGRS power spectra. Mon Not R Astron Soc 337:1068–1080. https://doi.org/10.1046/j.1365-8711.2002.06001.x

139. Anderson L, Aubourg E, Bailey S et al (2012) The clustering of galaxies in the SDSS-III baryon oscillation spectroscopic survey: baryon acoustic oscillations in the data release 9 spectroscopic galaxy sample. Mon Not R Astron Soc 427:3435–3467. https://doi.org/10.1111/j.1365-2966.2012.22066.x

140. Dawson KS, Schlegel DJ, Ahn CP et al (2013) The Baryon oscillation spectroscopic Survey of SDSS-III. Astron J 145:10. https://doi.org/10.1088/0004-6256/145/1/10

141. SDSS (2020) No need to Mind the Gap: Astrophysicists fill in 11 billion years of our universe's expansion history | SDSS | Press Releases. https://www.sdss.org/press-releases/no-need-to-mind-the-gap/. Accessed 20 Jul 2020

142. eBOSS Collaboration, Alam S, Aubert M et al (2020) The Completed SDSS-IV extended Baryon Oscillation Spectroscopic Survey: Cosmological Implications from two Decades of Spectroscopic Surveys at the Apache Point observatory. ArXiv E-Prints 2007. arXiv:2007.08991

6

The Age of Precision Cosmology

We began this book at the dawn of the twentieth century, when astronomers and physicists had only the vaguest understanding of space, time, matter, or the true size of the universe—much less the quantum nature of reality—and when speculations about the origin of the universe were widely considered "unscientific".

In the chapters that followed, we traced how both fields advanced their understanding through ever more precise measurements, profound theoretical insights, and often painful shifts in perspective. And we've arrived at today's era of precision cosmology, when physicists are hoping to achieve a fully unified theory of everything, astronomers are routinely measuring cosmological quantities within a percent or two [1], and the two fields have together converged on a *concordance* model that encompasses virtually all the available data.

This concordance is the lambda-cold-dark-matter (ΛCDM) model discussed in Chap. 5. But of course, no one expects the story to stop there. ΛCDM leaves too many loose ends. So in this final chapter, we'll look at three of those loose ends—and what they might tell us about what comes next.

© The Author(s), under exclusive license to Springer Nature
Switzerland AG 2022
M. M. Waldrop, *Cosmic Origins*,
https://doi.org/10.1007/978-3-030-98214-0_6

6.1   A Crisis Over the Age of the Universe?

The thing about living in an age of precision cosmology is that sometimes, you get two or more sets of very precise measurements that clash. Then what?

This is exactly what's been happening over the past decade with efforts to pin down H_0, the Hubble parameter that determines both the expansion rate and age of the universe.

Looking back on it, the evidence for this discrepancy had been accumulating for a while. But it didn't really start edging over into crisis territory until 2013, when the Planck satellite's science team released the first tranche of data from their observations of the cosmic microwave background (CMB) [2]. Included in that release was the group's best estimate for the Hubble parameter: 67.11 km per second per megaparsec, which pegged the age of the universe at 13.8 billion years. (Planck's current best figure, after further refinement, is 67.44 ± 0.58 in the same units [3]).

Getting that many significant digits in a cosmological number was an astonishing achievement, to say the least—except that two years earlier, the Supernovae, H_0 and Equation of State (SHoES) consortium based at the Space Telescope Science Institute in Baltimore had determined H_0 with just as much precision, but in a totally different way. First, the SHoES team had used the Hubble Space Telescope to monitor Cepheid variables and type 1a supernovae (SNe1a) in the same galaxies, which had allowed them to do a meticulous cross-calibration between these two different standard candles. And with those data in hand, they had arrived at a value for H_0 equal to 73.8 ± 2.4 km per second per megaparsec −10% higher than the Planck result [4]. (The current best value from the SHoES group is 73.30 ± 1.04 in the same units [5]). This would correspond to a universe that was expanding faster than the Planck measurement implied, and that was at least a billion years younger.

Now, discrepancies crop up all the time in astronomy, especially when it comes to cosmology. But a difference of more than four standard deviations? (That's a common measure of the estimated uncertainty). It wasn't something you could just chalk up to experimental uncertainty. So, in an effort to pinpoint where the problem lay, cosmologists turned to every alternative measure of the Hubble parameter that they could think of—and soon concluded that something was *really* wrong [6].

When they looked at values of H_0 derived from physics in the early universe, at or before the CMB formation at 380,000 years, they always got low numbers consistent with the Planck measurements. A good example was an estimate of about 68 for H_0 derived from a combination of baryon acoustic oscillations and primordial nucleosynthesis [7].

But when cosmologists looked at direct measures of H_0 based on data from the "late" universe—meaning, the recession velocities and distances to quasars, galaxies, clusters, and the like within a few billion light-years of the Milky Way—they always got high numbers consistent with the SHoES result. Examples included Cepheid variables in the Large Magellanic Cloud [8]; multi-image gravitational lenses [9]; and high-resolution tracking of naturally occurring "masers" as they orbit around supermassive black holes [10].

This "Hubble tension" has caused much consternation among cosmologists, and has inspired many attempts to resolve it. For example, skeptics of the early-universe numbers quickly noted that H_0 can't be determined from the CMB in isolation; it pops out of the data only when you take the ΛCDM model as a given—flat universe, cosmological constant, and all. Change that model, they argued, and maybe you could change the low H_0 number.

But change the model how? Theorists have struggled mightily to come up with a replacement that could resolve the Hubble tension while preserving ΛCDM's extraordinary successes, including its predictions for the flatness of the universe, the CMB anisotropies, the baryon acoustic oscillations, and all the rest. They've postulated the existence of new kinds of fields, new phases of matter that vanish just before the CMB forms—on and on. So far, however, none of these alternatives have gained wide acceptance [11].

On the other side, meanwhile, skeptics of the late-universe numbers noted that H_0 is just the ratio of an object's cosmic expansion velocity to its distance: $v = H_0 D$. Velocity can be measured quite accurately from the object's spectrum. But distance—well, that can be obtained only through an elaborate series of overlapping measurements known as the cosmic distance ladder. First, as discussed in Chap. 2, this ladder starts by using *parallax* to relate the known diameter of Earth's orbit to the distance of objects in our stellar neighborhood. Then these parallax measurements are used to calibrate the distance-luminosity relationship for various "standard candles" —notably the very bright Cepheid variable stars that can be seen in nearby galaxies. And finally, the Cepheids are used to calibrate an even brighter standard candle: the supernovae 1a explosions that allowed the discovery of cosmic acceleration. Thus the skepticism: When you do a string of overlapping calibrations like this, errors can accumulate. So maybe that was the source of all these late-universe, high-H_0 results. Somewhere along the line, someone had made a mistake.

But where? Astronomers have had lots of experience with the distance ladder by this point. The SHoES team, in particular, has done meticulous work, with multiple cross-checks; neither they nor anyone else has found a serious error. Furthermore, the late-universe distances obtained from techniques such as maser tracking are independent of the classic distance ladder—but yield a high H_0 anyway.

If nothing else, the impasse has put an intense focus on an alternative distance-ladder calibration method being explored by University of Chicago astronomer Wendy Freedman and her team. Known as the Tip of the Red Giant Branch (TRGB) technique, it calls for skipping over the Cepheid rung of the ladder entirely, and instead looking at the brightest red-giant stars in any given galaxy. Red giants represent an extremely luminous, end-of-life transition that stars like our Sun will go through as they are running out of fuel. And, because the physics of this transition is both well understood and quite consistent from one star to the next, the brightest red giants make for an excellent standard candle. TRGB stars can also be readily identified, even across millions of light years. On the near end, the red giants' intrinsic brightness can be calibrated through the use of parallax here in the Milky Way and Magellanic Clouds. And on the far end, they can be used to calibrate distances to the SNe1a.

In short, the red-giant technique promises to eliminate any subtle source of error lurking in the Cepheid calibration, and hopefully say which side of the Hubble tension was correct. But in September 2019, when Freedman and her team announced their new and highly anticipated value for H_0 using the TRGB technique, it came out to 69.8 ± 0.8—right in between the late-universe high values and the early-universe low values [12, 13].

As one astrophysicist tweeted when he heard the news, "The Universe is just messing with us at this point, right?" (Fig. 6.1).

In July 2020, two other much-anticipated surveys announced equally confounding results. On July 15 of that year, observers with the Atacama Cosmology Telescope in the Chilean Andes released the first ground-based map of CMB anisotropies that could rival the ones obtained by Planck [14]. Because ACT was independent of Planck, though, astronomers hoped it could provide a cross-check of the latter's H_0 value—and if need be, a correction. But ACT's H_0 came in at 67.9 ± 1.5 km per second per megaparsec, perfectly consistent with Planck.

Then two days later, on July 17, the Sloan Digital Sky Survey released its most massive compilation of galaxy redshifts ever—a 3D map of the heavens that used quasars and other tracers to fill in the 11 or so billion years between the CMB and the galaxies used in in the supernovae studies

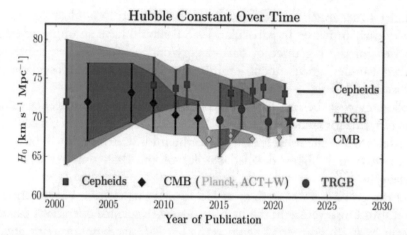

Fig. 6.1 Over the past decade or so, there has been an increasingly glaring discrepancy between values of the Hubble parameter, H_0, obtained from measurements in the "late" universe around our own galaxy (**Cepheids**), and in the "early" universe (**CMB**). A recently developed measure based on red giant stars (**TRGB**) is independent of the other two, and gives intermediate results. (Credit: Freedman WL (2021) Measurements of the Hubble Constant: Tensions in Perspective. The Astrophysical Journal 919:16. https://doi.org/10.3847/1538-4357/ac0e95, CC-BY 4.0)

[15]. Among other things, this survey used baryon acoustic oscillations and other data to confirm that the universe is flat to a high precision, and that dark energy has behaved like an unchanging constant over that entire span of time. And it yields an H_0 value of 68.20 ± 0.81, which is perfectly consistent with Planck's. But then, that might not be too surprising: In an attempt to avoid any uncertainties inherent in the traditional cosmic distance ladder, the SDSS team calibrated their velocity-distance relation with an "inverse distance ladder" that works backward from the CMB.

And so the tension simmers, with no resolution yet. The options are what they have been for the better part of a decade: Either there is something wrong with the ΛCDM model [8, 11], or there is something wrong with the distance ladder [16]. Or maybe Freedman is right, and the other two are both a little bit wrong—just in opposite directions.

Seeing with New Eyes

The good news, though, is that more and better data is on its way. On Christmas Day 2021, for example, the European Space Agency successfully launched the Hubble Space Telescopes successor: the James Webb Space Telescope (JWST). A joint project of NASA, ESA, and the Canadian Space Agency, JWST will have 10 times Hubble's light-gathering power, as well as an ability to see much further into the infrared spectrum—ideal for studying the very early universe. In particular, JWST should have an unprecedentedly clear view of the big three of dark-energy/dark-matter studies: supernovae standard candles, gravitational lensing, and baryon acoustic oscillations in the cosmic web.

Following close behind Webb will be the European Space Agency's Euclid spacecraft, due for launch in October 2022; and NASA's Nancy Grace Roman Space Telescope, scheduled for launch in the mid-2020s. Both will be specialized to carry out highly detailed studies of the dark universe at infrared wavelengths.

On the ground, a consortium of US funding agencies is constructing the Vera Rubin Observatory in the high Andes, with science operations expected to begin in 2023. Formerly known as the Large-Scale Synoptic Telescope, the observatory will photograph the entire visible sky in depth and detail—and then do it again, and again, and again, every few nights for a decade or more. Among other things, this enormous dataset will catalog billions of galaxies, give astronomers an unprecedented view of how dark matter influenced their formation, and reveal how (or if) dark energy has evolved over the lifetime of the universe.

And at microwave wavelengths, finally, a consortium of funding agencies and philanthropies is constructing the Simons Observatory at another site in the Andes—right next to the Atacama Cosmology Telescope, in fact. The Simons site will have several telescopes optimized for studying the CMB in even more detail than Planck could manage. One particular target will be *B-modes*: an extremely subtle, swirling pattern in the polarization of the CMB radiation. B-modes are widely considered to be the smoking-gun for inflation; finding them would be the first direct proof that inflation actually happened—and would finally allow astronomers to study exactly *how* it happened.

In sum, then, no one can predict when the new ideas will come. But with all this new data pouring in, no one should be surprised if the coming decade brings yet more revolutionary changes.

6.2   The Case of the Missing WIMPs

A second loose end is the question of what dark matter actually *is*. For decades now, the prevailing assumption has been that it's made of WIMPs: weakly interacting, massive particles left over from the Big Bang. But if that's the case, why hasn't anyone been able to detect them? Weakly interacting or not, shouldn't we have seen *something* by now?

In fact, physicists started dreaming up ways to detect the WIMPs almost as soon as the dark-matter-as-particles idea emerged in the mid-1980s. Their efforts since then have followed three major strategies—none of which has so far been successful [17–20].

Direct Detection

If the dark matter halo of the Milky Way really is made of weakly inter-acting particles, the reasoning goes, then zillions of them per second must be streaming through our solar system and everything in it—including us, the Earth, and any detector we can build. And, if these dark-matter particles are anything like neutrinos, then their interactions with normal matter will be weak, but not quite zero.

So this first strategy calls for researchers to take a target made of ordi-nary matter and put it somewhere deep underground where cosmic rays and other interference can't get at it. Then they just watch and wait: Eventually, if the above assumptions are correct, an atom in the target will spontaneously emit an explosive spray of particles as if it had been hit by something that came from nowhere [21]. That something, assuming that all the sources of interference and background noise can be eliminated, will be a dark-matter candidate.

The first proposal along these lines dates back to 1984 [22, 23]. The first actual experiments were carried out in 1986, with repurposed neutrino detec-tors that had target masses measured in kilograms [24, 25]. And many more experiments have followed, becoming larger, more sophisticated, and more sensitive over time; their current target masses are now measured in tons [17]. Along the way, these experiments have eliminated many theoretical possibil-ities for dark matter particles. Yet not one of them has seen an unambiguous dark matter signal.

In 2020, to take a recent example, the XENON1T team in Italy announced that their 3.3 tonne, liquid-xenon-filled detector in the deep underground Gran Sasso Laboratory had found a signal that was consis-tent with an axion coming from the sun, or possibly a new kind of neutrino—but that they could not rule out contamination from radioac-tive tritium[26]. Getting a more definitive answer will be a job for one of the next-generation dark-matter experiments now coming online. Prime examples include XENON1T's successor, the 8-tonne XENONnT in Gran Sasso; the 7-tonne, xenon-filled LUX-ZEPLIN (LZ) experiment located in an former gold mine in South Dakota; a next-generation xenon detector that will be built by the XENON and LZ groups jointly; the solid-state Super Cryogenic Dark Matter Search (SuperCDMS) at the SNOLab in Ontario; and the second-generation Axion Dark Matter eXperiment (ADMX G2) at the University of Washington. (The latter is the only experiment that does not need to be located deep underground. Instead, it will try to convert dark matter axions into observable microwaves using a strong magnetic field).

Astrophysical annihilation

The second strategy is to look up. If space is really filled with dark-matter particles, goes the reasoning, then pairs of them might occasionally meet and annihilate one another, converting their mass into detectable gamma rays [27, 28]. Or maybe they would annihilate into pairs of cosmic-ray protons and antiprotons [29]. Granted, such dark-matter annihilations would be exceedingly rare. But space is big, and it ought to contain a *lot* of WIMPs, so the total number of gamma rays, antiprotons, or whatever could be quite detectable. Alternatively, dark-matter particles trapped in the core of the Sun might accumulate over the eons and annihilate into neutrinos detectable here on Earth [30]. Either way, the energy and nature of the annihilation products could tell observers a lot about the original particles' mass and identity.

Again, however, despite major advances in our ability to detect gamma rays, cosmic rays, and astrophysical neutrinos, no one has yet found an unambiguous signal from dark-matter annihilation [17, 31]. Although it is still possible that some celestial antiprotons come from dark matter, for example, most (or all) are now thought to come from the collision of ordinary cosmic rays with interstellar gas. And although the center of the Milky Way does show a gamma-ray excess consistent with dark-matter annihilation [32, 33], conventional sources of gamma-ray signals are so abundant, and yet so poorly understood, that gamma-ray signals are notoriously hard to interpret.

New particle searches at the accelerators

The third strategy is to quit trying to detect dark matter particles out in the wild, so to speak, and instead make them in the lab—using, say, a high-energy accelerator such as the Large Hadron Collider (LHC) outside Geneva [34].

This is much easier said than done, though. Not only would the collision events that produced a dark-matter particle be exceeding rare, but the experimenters wouldn't actually see the particle once they'd made it: The thing would fly out of the detector (and the solar system) without interacting with anything else, as utterly invisible as its dark-matter siblings out in space. Instead, physicists would have to add up the energy, momentum, and spin of all the visible particles produced in the collision, then reconstruct the dark-matter particle's existence and properties by assuming that it had carried off everything that was missing. And even if they managed to pull that off, they would know only that they had discovered *an* invisible particle, not that it was *the* invisible dark-matter particle.

Still, this kind of unsee-able particle signal would not be so different from the events produced by neutrinos, and physicists have had lots of practice at working with them. Accelerators would undoubtedly be the best place to study dark-matter particles when and if they were discovered by either of the first two methods. And, because the existence of dark matter is the most definitive hard evidence that there is some kind of physics beyond the standard model, physicists have been eager to get started that study.

All of which is why it's so worrisome that neither the LHC nor any other accelerator has yet found the slightest hint of a dark-matter particle. And that's only one piece of a much larger concern. The standard model of particle physics is now complete, thanks to the LHC's spectacular discovery of the long-predicted Higgs Boson in 2012 (see Chap. 4). But since then, the LHC hasn't found anything that's *not* in the standard model: no supersymmetry, no extra dimensions, no dark-matter candidates, nothing new at all. Of course, that could all change tomorrow—or more precisely, at some point in 2022, when the LHC is scheduled to finish its current round of upgrades and start generating data again with its beams set to a much higher intensity. This should allow it to do a much more thorough search for rare events. Likewise, events could start appearing one of the new-generation dark matter detectors mentioned above.

But no matter what happens, the previous searches have already ruled out most of the natural and attractive extensions of the standard model—and with them, most of the WIMP candidates [20, 35, 36]. Axions are still very possible. But even so, the non-appearance of WIMPs has forced theorists of all stripes back to their proverbial drawing boards in search of alternatives—any of which could require a rethink of how the physics of particles shaped the course of the Big Bang.

One intriguing possibility that's getting a fresh look is that dark matter consists of primordial black holes: tiny gravitational singularities produced during the Big Bang [37]. Theorists disagree on exactly how (or whether) enough black holes could have formed in those initial instants; the calculations are messy and difficult. But they would not count as baryonic matter, so they would not be subject to the abundance limits discussed in Chap. 5. And observers can't rule them out. The latest data suggest that black holes

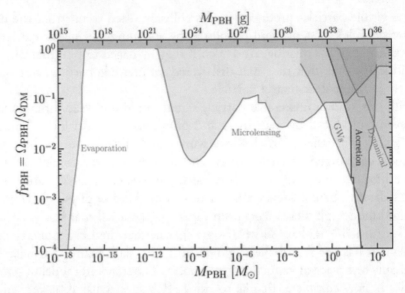

Fig. 6.2 The latest observations show that asteroid-mass primordial black holes (PBHs) with masses between 10^{17} and 10^{22} g remain viable candidates for dark matter. Lower-mass PBHs would have evaporated already through a process known as Hawking radiation, while their higher-mass cousins are ruled out by the microlensing observations discussed in Chap. 5, by gravitational wave (GW) observations, and by other astrophysical constraints. (Credit: Green AM, Kavanagh BJ (2021) Primordial black holes as a dark matter candidate. J Phys G: Nucl Part Phys 48:043,001. https://doi.org/10.1088/1361-6471/abc534, CC-BY 4.0)

with the mass of asteroids—between 10^{17} and 10^{22} g—would have escaped detection so far (Fig. 6.2). So they continue to be viable possibilities.[1]

And then there is the hypothetical "hidden sector"—a whole new realm of particles that would have rich interactions with one another, but only affect standard-model particles through gravity [20, 38]. The latter wouldn't show up in dark-matter detectors or accelerators, but might still produce detectable annihilation events in the sky [39].

On the experimental front, meanwhile, a number of researchers have pointed out that most existing detectors assume that dark matter particles are heavier than the proton—and would have missed them if the particles are actually very light. So instead, these scientists have proposed detectors in which light dark-matter particles passing through a solid-state device could

---

[1] Fortunately, the odds of such a dark-matter black hole striking the Earth are infinitesimal. And we might not notice even if one did (depending on who or what it hit). A black hole in that mass range would be the size of a hydrogen atom or smaller, and moving at several hundred kilometers per second; it would probably drill a *very* thin tunnel all the way through the planet and out the other side, without noticeably slowing down.

excite a subtle electronic excitation known as a plasmon [40, 41], or an even more subtle spin excitation known as a magnon [42].

In short, the near future is rife with possibilities for dark-matter detection—or not. So stay tuned.

6.3   Concordance—and Beyond?

A final loose end is the ΛCDM framework itself. It's called a concordance model because it's supposed to explain all the observations in the simplest terms possible, but no simpler.[2] And it does—which is precisely the problem: Those simplest possible terms are still pretty complicated. Depending on how you count, the ΛCDM account requires at least six key elements. We've covered each of them in earlier chapters, but to recap:

1. **General Relativity**

   Developed by Albert Einstein in 1915, this theory explains how the force of gravity arises from the curvature of space and time. And, as Einstein and others soon realized, it provides a theoretical framework for understanding the evolution of the universe (see Chap. 2).

2. **The Standard Model of Particle Physics**

   Developed by multiple physicists from the 1950s through the 1970s, this theory provides a unified account of electricity, magnetism, light, the strong nuclear force, and the weak force responsible for certain forms of radioactivity—that is, all the known forces *except* gravity (see Chap. 4).

3. **The Big Bang**

   Our universe originated in a hot, dense, explosive event some 13.8 billion years ago, and has been expanding outward ever since. Although the notion of a cosmic beginning was controversial early on, three observations made it almost inescapable:

   - **Cosmic expansion**, which is predicted by general relativity and confirmed every time astronomers look at a far-off galaxy and see it receding with a redshift proportional to its distance (see Chap. 2).
   - **Big Bang nucleosynthesis**, which took place during the first few minutes of the expansion, as the primordial plasma that filled the

---

[2] Although this quote—"Everything should be made as simple as possible, but not simpler"—is often attributed to Einstein, there is <u>no evidence</u> that he ever said it. Still, he made plenty of comments along the same lines, and it's clear that he would have agreed with the sentiment.

infant universe transformed itself into a mix of hydrogen, deuterium, and helium isotopes. When this era is modeled using modern nuclear physics, the calculations show that these isotopes would have been produced in ratios that very closely match what we see in the universe today (see Chaps. 3 and 4).

- **The Cosmic Microwave Background (CMB)**, which is the Big Bang's afterglow, redshifted by 13.8 billion years of cosmic expansion. We see it today as a faint whisper of microwaves streaming down from every direction in the sky—a signal that's obvious to anyone with a receiver tuned to the right frequencies (see Chap. 3).

4. Inflation

The Big Bang itself emerged from a period of exponential expansion that was both inconceivably brief and unimaginably rapid. This expansion inflated the observable universe like a cosmic balloon, stretching it out until space was virtually flat. But inflation also left behind a series of quantum-scale density fluctuations that would eventually grow into today's cosmic web of galaxies. It would be too much to say that the concordance model also includes the theory of eternal inflation and the multiverse, which holds that our Big Bang was only one of many; this idea is too hard to prove observationally, and leads to anthropic arguments that too many researchers find distasteful. But if you accept inflation at all, the multiverse difficult to avoid (see Chap. 4).

5. Dark Matter

The universe is partially filled with a mysterious, utterly invisible substance that outweighs the ordinary, baryonic matter that we are made of by something like 5 to 1, and exerts a gravitational pull strong enough to dominate the dynamics and formation of visible galaxies. No one can be completely sure what dark matter is, but it behaves like a swarm of weakly interacting, massive particles (WIMPs) left over from the Big Bang. About the only thing anyone *can* say for sure, based on computer simulations, is that the particles must be "cold"—that is, moving much slower than the speed of light. Thus the "cold dark matter" part of the concordance name (see Chap. 5).

6. Dark Energy

The cosmic expansion is slowly being accelerated by an equally invisible, but even more mysterious substance that accounts for roughly 70% of all the mass/energy in the universe. Dark energy doesn't seem to be made of particles, weakly interacting or otherwise; as far as anyone can tell, it affects every point of space-time equally, and is mathematically equivalent to the cosmological constant that Einstein proposed back in 1917. Following Einstein's lead, cosmologists ever since have inserted this constant into their formulas using the Greek letter Λ, pronounced *lambda*. Thus the ΛCDM name (see Chaps. 2 and 5).

The problem with this list isn't so much its length (although half a dozen moving parts *is* a lot), but the fact that so many of the pieces are just … *there*, with no explanation. This is particularly true of the last three items: inflation, dark matter, and dark energy. Each of them requires some kind of physics that goes beyond the standard model. Yet none of them bears any obvious relation to the others.

Worse, the ΛCDM concordance leaves a lot of unanswered questions—starting with the simplest and most fundamental: "What banged?" Or to put in a slightly more formal way, how did dark energy, dark matter, inflation, and the Big Bang arise from physical law? And how do they each fit in with the multiverse and the anthropic principle—if indeed they do?

Getting a truly satisfactory answer to such questions will almost certainly have to wait for that elusive Theory of Everything—the long-sought, full-on unification of quantum mechanics, general relativity, the standard model, and who knows what else.

Theorists have come up with plenty of candidates for such a unification, some of which we discussed in Chap. 4. One is superstring theory, which posits that the lumps we perceive as particles are actually infinitesimal, madly vibrating threads of energy. Since its emergence in the 1970s, superstring theory has become by far the most popular candidate for a Theory of Everything, not least because it can give a mathematically elegant explanation for the existence of both gravity, and the forces between particles [43]. But Chap. 4 also discussed superstring theory's strongest contender, Loop Quantum Gravity (LQG). Basically, LQG starts by rewriting the equations of Einstein's general relativity in a form that's (comparatively) easy to quantize. And it ends up showing how space and time themselves could emerge from elementary quanta of area and volume, in much the same way that liquid water or solid ice emerge from huge numbers of H_2O molecules [44–48].

As we also saw in Chap. 4, both these approaches have been used to model cosmic origins, whether through the superstring landscape [49–52] or Loop Quantum Cosmology [53–59]. But alas, neither of them quite makes it to Theory of Everything status. Superstrings come closest, but no one has yet been able to explain how they could give rise to the precise set of forces and particles seen in the standard model. And LQG doesn't even talk about particles and forces; in its basic form, at least, it's a quantum theory of space, time, and gravity alone.

So theorists continue to look elsewhere. One particularly intriguing approach that's emerged over the past two decades is a conjecture known as *gauge-gravity duality* [60, 61]. It holds that gravity on a curved space–time can be described by general relativity, *or* by a more-or-less normal quantum field theory operating on a flat space–time that has one less spatial dimension—with no gravity in sight. Granted, this has never been proved rigorously. And at least in its original form, the relation only connected gravity in a type of universe known as *anti-de Sitter space*, to a *conformal field theory* on the boundary of that space.[3] While each of them bears many similarities to the universe we live in, they also have some profound differences. Still, within those limits the duality has passed innumerable consistency checks, and is widely believed to be true—meaning that the two ways of describing the same reality are mathematically equivalent, and give absolutely identical predictions.

And that assumption, in turn, has given theorists an extraordinarily rich tool for exploring the relationship between gravity and the quantum realm. Perhaps the most provocative possibility coming out of this work is that the very fabric of space–time emerges from quantum *entanglement*: a (very common) situation in which particles have their quantum states intertwined in a way that's independent of distance [62]. Among other things, this possibility suggests that two particles connected by quantum entanglement may somehow be equivalent to two black holes connected by a space–time tunnel called an Einstein-Rosen bridge—or as it's more popularly known, a *wormhole* [63].

Even if true, however, neither the gauge-gravity duality nor the entanglement-wormhole equivalence is remotely close to being a complete Theory of Everything. So again, there is a widespread feeling among researchers in this field that still more new ideas are needed.

Good ideas of this magnitude are rare and unpredictable, to put it mildly. But not impossible: there's always the example of 1905, after all, when a lowly

---

[3] Thus the other common name for this approach: the AdS-CFT duality. Anti-de Sitter space describes an eternally expanding universe that contains no matter, but that does have a negative cosmological constant. It's that last part that makes AdS space so different from our own, which is much closer to the cosmological solution found by the Dutch physicist Willem de Sitter in 1917—namely, an empty, expanding universe with a *positive* cosmological constant. One paradoxical property of AdS is that it is infinite, yet has a boundary. This boundary is geometrically flat, and is where the conformal field theory lives. This quantum field theory is similar in many ways to its cousins that describe electrons, quarks, and all the other particles in the standard model. But it also differs in some critical ways—not the least being that it can only describe massless particles.

examiner in the Swiss patent office managed to revolutionize all of modern physics in the single year. So once again, stay tuned.

References

1. Turner MS (2022) The road to precision cosmology
2. Collaboration P, Ade PAR, Aghanim N et al (2014) Planck 2013 results. XVI. Cosmological parameters. Astron Astrophys 571:A16. https://doi.org/10.1051/0004-6361/201321591
3. Planck Collaboration, Aghanim N, Akrami Y, et al (2018) Planck 2018 results. VI. Cosmological parameters. ArXiv E-Prints 1807:arXiv:1807.06209
4. Riess AG, Macri L, Casertano S et al (2011) A 3% solution: determination of the hubble constant with the hubble space telescope and wide field camera 3. Astrophys J 730:119. https://doi.org/10.1088/0004-637X/730/2/119
5. Riess AG, Yuan W, Macri LM, et al (2021) A comprehensive measurement of the local value of the hubble constant with 1 km/s/Mpc Uncertainty from the hubble space telescope and the SH0ES Team. ArXiv E-Prints
6. Verde L, Treu T, Riess AG (2019) Tensions between the early and late Universe. Nat Astron 3:891–895. https://doi.org/10.1038/s41550-019-0902-0
7. Cuceu A, Farr J, Lemos P, Font-Ribera A (2019) Baryon Acoustic Oscillations and the Hubble constant: past, present and future. J Cosmol Astropart Phys 10:044. https://doi.org/10.1088/1475-7516/2019/10/044
8. Riess AG, Casertano S, Yuan W et al (2019) Large magellanic cloud cepheid standards provide a 1% foundation for the determination of the hubble constant and stronger evidence for physics beyond ΛCDM. Astrophys J 876:85. https://doi.org/10.3847/1538-4357/ab1422
9. Shajib AJ, Birrer S, Treu T et al (2020) STRIDES: a 3.9 per cent measurement of the Hubble constant from the strong lens system DES J0408–5354. Mon Not R Astron Soc 494:6072–6102. https://doi.org/10.1093/mnras/staa828
10. Pesce DW, Braatz JA, Reid MJ, et al (2020) The Megamaser cosmology project. XIII. Combined hubble constant constraints. Astrophys J Lett 891:L1. https://doi.org/10.3847/2041-8213/ab75f0
11. Knox L, Millea M (2020) Hubble constant hunter's guide. Phys Rev D 101:043533.https://doi.org/10.1103/PhysRevD.101.043533
12. Freedman WL, Madore BF, Hatt D, et al (2019) The Carnegie-Chicago hubble program. VIII. An independent determination of the hubble constant based on the tip of the red giant branch. Astrophys J 882:34. https://doi.org/10.3847/1538-4357/ab2f73
13. Freedman WL, Madore BF, Hoyt T et al (2020) Calibration of the tip of the red giant branch. Astrophys J 891:57. https://doi.org/10.3847/1538-4357/ab7339

14. Aiola S, Calabrese E, Maurin L et al (2020) The atacama cosmology telescope: DR4 maps and cosmological parameters. ArXiv E-Prints 2007:arXiv:2007.07288

15. eBOSS Collaboration, Alam S, Aubert M et al (2020) The completed SDSS-IV extended baryon oscillation spectroscopic survey: cosmological implications from two decades of spectroscopic surveys at the apache point observatory. ArXiv E-Prints 2007:arXiv:2007.08991

16. Efstathiou G (2020) A lockdown perspective on the hubble tension. ArXiv E-Prints 2007:arXiv:2007.10716

17. Bertone G, Hooper D (2018) History of dark matter. Rev Mod Phys 90:045002.https://doi.org/10.1103/RevModPhys.90.045002

18. Bertone G, Hooper D, Silk J (2005) Particle dark matter: evidence, candidates and constraints. Phys Rep 405:279–390. https://doi.org/10.1016/j.phy srep.2004.08.031

19. Hooper D, Baltz EA (2008) Strategies for determining the nature of dark matter. Annu Rev Nucl Part Sci 58:293–314. https://doi.org/10.1146/annurev. nucl.58.110707.171217

20. Bertone G, Tait TMP (2018) A new era in the search for dark matter. Nature 562:51–56. https://doi.org/10.1038/s41586-018-0542-z

21. Gaitskell RJ (2004) Direct detection of dark matter. Annu Rev Nucl Part Sci 54:315–359. https://doi.org/10.1146/annurev.nucl.54.070103.181244

22. Drukier A, Stodolsky L (1984) Principles and applications of a neutral-current detector for neutrino physics and astronomy. Phys Rev D 30:2295–2309. https://doi.org/10.1103/PhysRevD.30.2295

23. Goodman MW, Witten E (1985) Detectability of certain dark-matter candidates. Phys Rev D 31:3059–3063. https://doi.org/10.1103/PhysRevD.31.3059

24. Ahlen SP, Avignone FT, Brodzinski RL et al (1987) Limits on cold dark matter candidates from an ultralow background germanium spectrometer. Phys Lett B 195:603–608. https://doi.org/10.1016/0370-2693(87)91581-4

25. Caldwell DO, Eisberg RM, Grumm DM et al (1988) Laboratory limits on Galactic cold dark matter. Phys Rev Lett 61:510–513. https://doi.org/10.1103/PhysRevLett.61.510

26. Aprile E, Aalbers J, Agostini F et al (2020) Observation of excess electronic recoil events in XENON1T. ArXiv E-Prints 2006:arXiv:2006.09721

27. Gunn JE, Lee BW, Lerche I et al (1978) Some astrophysical consequences of the existence of a heavy stable neutral lepton. Astrophys J 223:1015–1031. https://doi.org/10.1086/156335

28. Stecker FW (1978) The cosmic gamma-ray background from the annihilation of primordial stable neutral heavy leptons. Astrophys J 223:1032–1036. https://doi.org/10.1086/156336

29. Silk J, Srednicki M (1984) Cosmic-ray antiprotons as a probe of a photino-dominated universe. Phys Rev Lett 53:624–627. https://doi.org/10.1103/PhysRevLett.53.624

30. Krauss LM, Freese K, Spergel DN, Press WH (1985) Cold dark matter candidates and the solar neutrino problem. Astrophys J 299:1001–1006. https://doi.org/10.1086/163767

31. Porter TA, Johnson RP, Graham PW (2011) Dark matter searches with Astroparticle data. Annu Rev Astron Astrophys 49:155–194. https://doi.org/10.1146/annurev-astro-081710-102528

32. Goodenough L, Hooper D (2009) Possible evidence for dark matter annihilation in the inner milky way from the fermi gamma ray space telescope. ArXiv E-Prints 0910:arXiv:0910.2998

33. Daylan T, Finkbeiner DP, Hooper D et al (2016) The characterization of the gamma-ray signal from the central Milky Way: a case for annihilating dark matter. Phys Dark Universe 12:1–23. https://doi.org/10.1016/j.dark.2015.12.005

34. Boveia A, Doglioni C (2018) Dark matter searches at colliders. Annu Rev Nucl Part Sci 68:429–459. https://doi.org/10.1146/annurev-nucl-101917-021008

35. Feng JL (2013) Naturalness and the status of supersymmetry. Annu Rev Nucl Part Sci 63:351–382. https://doi.org/10.1146/annurev-nucl-102010-130447

36. Dine M (2015) Naturalness Under Stress. Annu Rev Nucl Part Sci 65:43–62. https://doi.org/10.1146/annurev-nucl-102014-022053

37. Green AM, Kavanagh BJ (2021) Primordial black holes as a dark matter candidate. J Phys G Nucl Part Phys 48:043001.https://doi.org/10.1088/1361-6471/abc534

38. Feng JL (2010) Dark matter candidates from particle physics and methods of detection. Annu Rev Astron Astrophys 48:495–545. https://doi.org/10.1146/annurev-astro-082708-101659

39. Hooper D, Leane RK, Tsai Y-D et al (2019) A systematic study of hidden sector dark matter: application to the gamma-ray and antiproton excesses. ArXiv191208821 Hep-Ph

40. Kurinsky N, Baxter D, Kahn Y, Krnjaic G (2020) A dark matter interpretation of excesses in multiple direct detection experiments. ArXiv E-Prints 2002:arXiv:2002.06937

41. Kozaczuk J, Lin T (2020) Plasmon production from dark matter scattering. Phys Rev D 101:123012.https://doi.org/10.1103/PhysRevD.101.123012

42. Trickle T, Zhang Z, Zurek KM (2020) Detecting Light Dark Matter with Magnons. Phys Rev Lett 124:201801.https://doi.org/10.1103/PhysRevLett.124.201801

43. Polchinski J (2005) String theory

44. Rovelli C, Smolin L (1988) Knot theory and quantum gravity. Phys Rev Lett 61:1155–1158. https://doi.org/10.1103/PhysRevLett.61.1155

45. Rovelli C, Smolin L (1990) Loop space representation of quantum general relativity. Nucl Phys B 331:80–152. https://doi.org/10.1016/0550-3213(90)90019-A

46. Ashtekar A, Rovelli C, Smolin L (1992) Weaving a classical metric with quantum threads. Phys Rev Lett 69:237–240. https://doi.org/10.1103/PhysRevLett.69.237

47. Rovelli C, Smolin L (1995) Spin networks and quantum gravity. Phys Rev D 52:5743–5759. https://doi.org/10.1103/PhysRevD.52.5743

48. Rovelli C, Smolin L (1995) Discreteness of area and volume in quantum gravity. Nucl Phys B 442:593–619. https://doi.org/10.1016/0550-3213(95)00150-Q

49. Bousso R, Polchinski J (2000) Quantization of four-form fluxes and dynamical neutralization of the cosmological constant. J High Energy Phys 06:006. https://doi.org/10.1088/1126-6708/2000/06/006

50. Douglas MR (2003) The statistics of string/M theory vacua. J High Energy Phys 05:046. https://doi.org/10.1088/1126-6708/2003/05/046

51. Susskind L (2003) The anthropic landscape of string theory, 26

52. Kachru S, Kallosh R, Linde A et al (2003) Towards inflation in string theory. J Cosmol Astropart Phys 10:013. https://doi.org/10.1088/1475-7516/2003/10/013

53. Bojowald M (2000) Loop quantum cosmology: I. Kinematics. Class Quantum Gravity 17:1489–1508. https://doi.org/10.1088/0264-9381/17/6/312

54. Bojowald M (2000) Loop quantum cosmology: II. Volume operators. Class Quantum Gravity 17:1509–1526. https://doi.org/10.1088/0264-9381/17/6/313

55. Bojowald M (2001) Loop quantum cosmology: III. Wheeler-DeWitt operators. Class Quantum Gravity 18:1055–1069. https://doi.org/10.1088/0264-9381/18/6/307

56. Bojowald M (2001) Loop quantum cosmology: IV. Discrete time evolution. Class Quantum Gravity 18:1071–1087. https://doi.org/10.1088/0264-9381/18/6/308

57. Bojowald M (2001) Absence of a singularity in loop quantum cosmology. Phys Rev Lett 86:5227–5230. https://doi.org/10.1103/PhysRevLett.86.5227

58. Bojowald M (2008) Loop quantum cosmology. Living Rev Relativ 11:4. https://doi.org/10.12942/lrr-2008-4

59. Ashtekar A, Singh P (2011) Loop quantum cosmology: a status report. Class Quantum Gravity 28:213001.https://doi.org/10.1088/0264-9381/28/21/213001

60. Maldacena JM (1998) The Large N limit of Superconformal field theories and supergravity. Adv Theor Math Phys 2:231

61. Maldacena J (2012) The gauge gravity duality. In: Proceedings of Xth quark confinement and the hadron spectrum. TUM Camous Garching, Munich, Germany, pp 8–12

62. Van Raamsdonk M (2010) Building up spacetime with quantum entanglement. Int J Mod Phys D 19:2429–2435. https://doi.org/10.1142/S0218271810018529

63. Maldacena J, Susskind L (2013) Cool horizons for entangled black holes. Fortschritte Phys 61:781–811. https://doi.org/10.1002/prop.201300020

Printed in the United States
by Baker & Taylor Publisher Services